T0140750

Reactions of Hydroxyl Radicals with Oxygenated Hydrocarbons in the Gas Phase: A Laser Photolysis/Laser-Induced Fluorescence Study

Zur Erlangung des akademischen Grades einer

DOKTORIN DER NATURWISSENSCHAFTEN

(Dr. rer. nat.)

der KIT-Fakultät für Chemie und Biowissenschaften
des Karlsruher Institut für Technologie (KIT)
genehmigte

DISSERTATION

von

Dipl.-Chem. Cornelie Bänsch
geb. Hüllemann

aus Neubulach-Oberhaugstett

Dekan: Prof. Dr. R. Fischer
Referent: Prof. Dr. M. Olzmann
Korreferent: Prof. Dr. M. Kappes
Tag der mündlichen Prüfung: 16.10.2017

Bibliographic information published by the Deutsche Nationalbibliothek

The Deutsche Nationalbibliothek lists this publication in the Deutsche Nationalbibliografie; detailed bibliographic data are available on the Internet at http://dnb.d-nb.de .

ISBN 978-3-8325-4632-8

Logos Verlag Berlin GmbH
Comeniushof, Gubener Str. 47,
10243 Berlin
Tel.: +49 (0)30 42 85 10 90
Fax: +49 (0)30 42 85 10 92
INTERNET: http://www.logos-verlag.de

Acknowledgement

An dieser Stelle möchte ich mich bei allen Menschen bedanken, die das Gelingen dieser Arbeit möglich gemacht haben.

An erster Stelle geht mein aufrichtiger Dank an meinen Doktorvater, Prof. Dr. Matthias Olzmann, der mir die Bearbeitung der interessanten Fragestellungen ermöglichte und mir jederzeit mit konstruktiven und kreativen Ratschlägen zur Seite stand. Vielen Dank für die gewährten wissenschaftlichen Freiräume und für das in vielerlei Hinsicht entgegengebrachte Vertrauen.

Mein ganz besonderer Dank gilt meinen langjährigen Laborkollegen Dr. Jens Hetzler und Julia Eble für die tolle Zusammenarbeit, die entspannte Atmosphäre und die vielen hilfreichen Diskussionen. Außerdem möchte ich mich bei Dr. Gabor Zügner für die gute Abstimmung und die sehr sorgfältige Weiterführung des Labors in meiner Abwesenheit bedanken.

Sehr herzlich möchte ich mich auch bei meinen Kollegen Dr. Johannes Kiecherer, Dr. Mark Pfeifle und Dr. Milan Szöri für die spannende Zusammenarbeit in den einzelnen Projekten und ihre unermüdliche Geduld, mir "theoretisch" weiterzuhelfen, bedanken.

Ein aufrichtiger Dank geht auch an unsere technische Mitarbeiterin Patricia Hibomvschi. Ohne ihre Hilfe in den zahlreichen technischen Problemen, die sich während der Arbeit ergeben haben, wäre ein Gelingen nicht möglich gewesen. Zusätzlich haben mir die vielen tollen Gespräche jeglicher Natur stets den Arbeitsalltag erleichtert.

Weiterhin möchte ich mich herzlich bei der mechanischen Werkstatt, insbesondere bei Dieter Walz, Fritz Siegel und Thorsten Franzke, und der Elektrowerkstatt mit Klaus Stree und Holger Halberstadt bedanken. Sie waren immer für meine Ideen in technischer Hinsicht offen und stets in kleinen und großen Notfällen zur Stelle.

Mein Dank gilt auch unserer Sekretärin Daniela Rohmert-Hug für die Unterstützung in den unzähligen organisatorischen Fragen.

Außerdem möchte ich mich bei der gesamten Arbeitsgruppe und meinen vielen ehemaligen Kollegen bedanken. Vielen Dank für die lockere und tolle Atmosphäre, die gegenseitige Unterstützung, das große hervorgebrachte Verständnis und die zahlreichen schönen gemeinsam verbrachten Stunden. An Desmond, Julia, Dennis, Christian, Leonie und Corina auch ein herzliches Dankeschön für das Korrekturlesen dieser Arbeit.

Zuletzt gilt mein Dank meinen Freunden und meiner Familie. Vielen Dank für die zahlreichen Gespräche, die Ermutigungen und den nötigen Ausgleich, den ihr mir gegeben habt.

Mein besonderer Dank geht an meine Eltern. Ihr wart immer für mich da und habt mir stets liebevoll und offen zur Seite gestanden.

Am Ende möchte ich mich noch bei meinen zwei Liebsten, meinem Mann Johannes und meiner Tochter Paula, bedanken. Durch euch habe ich immer wieder Kraft, Mut und Zuversicht gefunden und nie den Glauben daran verloren, Hürden und Herausforderungen meistern zu können. Euch an meiner Seite zu haben, ist mein größtes Glück.

Contents

Abstract

Hydroxyl radicals (OH) play an important role in gas phase chemistry. Especially for the modeling of combustion and atmospheric processes, the detailed knowledge of the kinetics of OH reactions in a broad range of conditions is of fundamental importance. In this work, an experimental approach for time-resolved studies of OH radical reactions at pressures above 1 bar with pulsed laser photolysis/laser-induced fluorescence (PLP/LIF) was revised, and several OH reaction systems were investigated.

In the first part, preliminary studies with UV/Vis absorption spectroscopy on the purity of nitric acid (HNO_3), which was used as the OH precursor, are presented and the resulting revision of the PLP/LIF setup is explained. With the absorption measurements, nitrogen dioxide (NO_2) was identified as the most important impurity in the hitherto used HNO_3 gas mixtures. It could be shown that systematic errors in the kinetic studies can arise from this impurity when the original PLP/LIF setup is applied. Consequently, the setup was modified in a manner that an application of gas mixtures is possible without causing systematic errors. Moreover, a method was developed which enables the supply of gaseous nitric acid on high levels of purity at bath gas pressures above 1 bar. The centerpiece of this setup is a bubbler, which can be used at bath gas pressures of up to 100 bar. In subsequent PLP/LIF measurements, the developed experimental approach for the supply of pure gaseous nitric acid was tested. To this end, the rate coefficient of the reaction

$$\mathrm{HNO_3 + OH \xrightarrow{k_1} products} \tag{R1}$$

was investigated at 296 K and 10 bar. A value of $k_1 = (1.6\pm0.2)\ 10^{-13}\ \mathrm{cm^3\ s^{-1}}$ was obtained, which is in very good agreement with the proposed high-pressure limit from the literature.

The second part of this work is concerned with the kinetics of the reactions of OH radicals with dimethyl ether (DME), diethyl ether (DEE), dimethoxymethane (DMM), and some of their isotopologues:

$$\mathrm{DME(-d_6) + OH \xrightarrow{k_{2(d)}} products} \tag{R2}$$

$$\mathrm{DEE(-d_{10}) + OH \xrightarrow{k_{3(d)}} products} \tag{R3}$$

$$\mathrm{DMM + OH \xrightarrow{k_4} products.} \tag{R4}$$

The kinetics of the reaction DME + OH was studied in the prior diploma thesis of the author. Hence, only the rate coefficient k_{2d}, which corresponds to the reaction of the perdeuterated isotopologe of DME, was investigated in this work at temperatures between 387 and 554 K in a pressure range between 13.0 and 20.4 bar. No pressure dependence of k_{2d} was observed under these conditions. A positive temperature dependence with a marginal curvature in the Arrhenius behavior was obtained. The following expression resulted as best fit:

$$k_{2d} = 7.27 \cdot 10^{-23} \left(\frac{T}{\mathrm{K}}\right)^{3.568} \exp\left(\frac{780\ \mathrm{K}}{T}\right)\ \mathrm{cm^3\ s^{-1}}. \tag{0.1}$$

The experimentally determined isotope effect k_2/k_{2d} amounts to approx. 3.5 in the considered temperature range. A deviation of the measured k_{2d} from literature values by a factor of two was observed. However, theoretical studies of the isotope effect from this group, and from the literature, support the results obtained in this work. By comparing the experimental findings with these studies, the suitability of quantum chemical calculations at the CCSD(T)/cc-pV(T+Q)Z//CCSD/cc-pVDZ level of theory for the description of such type of systems could be verified, while the results obtained at CBS-QB3 level of theory could not be confirmed.

In the study of the system DEE + OH, both k_3 and k_{3d} were determined experimentally in a temperature range between 295 and 570 K at 2, 5, and 10 bar pressure. No dependence on the pressure resulted for both rate coefficients. For k_3 a good agreement between the the direct measurements from the literature and the results from this work was obtained. Altogether, a curved Arrhenius behavior with a minimum at around 400 K was observed. As a result, an overall fit was conducted yielding the following expression:

$$k_3 = 1.46 \cdot 10^{-17} \left(\frac{T}{\mathrm{K}}\right)^{1.948} \exp\left(\frac{778\ \mathrm{K}}{T}\right)\ \mathrm{cm^3\ s^{-1}}. \tag{0.2}$$

A positive temperature dependence with a slight curvature of the Arrhenius plot resulted for k_{3d}. The best fit is given by

$$k_{3d} = 9.82 \cdot 10^{-16} \left(\frac{T}{\mathrm{K}}\right)^{1.334} \exp\left(\frac{343\ \mathrm{K}}{T}\right)\ \mathrm{cm^3\ s^{-1}}. \tag{0.3}$$

Hence, an isotope effect k_3/k_{3d} of approx. 2 was obtained from the experiment under the considered conditions. The results were compared to calculations of $k_{3(d)}$ from this group, and the general theoretical description of the reaction could be validated.

The measurements of the rate coefficient k_4 of the reaction DMM + OH were carried out at temperatures between 297 and 570 K at 2, 5, and 10 bar pressure. Under consideration of the experimental error no pressure dependence of k_4 was deduced. A negative temperature dependence with no significant curvature in the Arrhenius behavior was observed. Thus, the values could be fitted adequately with an expression described by

$$k_4 = 2.90 \cdot 10^{-12} \exp\left(\frac{126\ \mathrm{K}}{T}\right)\ \mathrm{cm^3\ s^{-1}}\,. \tag{0.4}$$

Up to 370 K, the results agree well with measurements from the literature. At higher temperatures a divergence between the rate coefficients determined in this work and the results of the only study from the literature under these conditions was observed. The cause of this divergent behavior remains unclear. The mechanistic implications of the differing experimental findings and the consequences for the modeling of ignition processes of DMM are discussed.

Zusammenfassung

Reaktionen von Hydroxylradikalen (OH) sind von großer Relevanz für die Atmosphären- und Verbrennungschemie. Die vorliegende Arbeit beschäftigt sich mit der experimentellen Untersuchung von OH-Reaktionen in der Gasphase mit gepulster Laserphotolyse und laserinduzierter Fluoreszenz (PLP/LIF) unter Hochdruckbedingungen.

Im ersten Teil werden vorbereitende Untersuchungen mittels Absorptionspektroskopie zur Reinheit von Salpetersäure (HNO_3), die als OH-Vorläufer verwendet wurde, vorgestellt und die daraus resultierende Überarbeitung des PLP/LIF-Aufbaus erklärt. Im Rahmen der Absorptionsmessungen wurde Stickstoffdioxid (NO_2) als hauptsächliche Verunreinigung der bisher verwendeten Salpetersäure-Gasmischungen identifiziert. Es konnte gezeigt werden, dass durch diese Verunreinigung systematische Fehler in den kinetischen Untersuchungen entstehen können, wenn der bisherige PLP/LIF-Aufbau verwendet wird. Infolgedessen wurde der Aufbau so modifiziert, dass die Verwendung von Gasmischungen ohne systematische Fehler möglich ist. Des Weiteren wurde ein Verfahren entwickelt, das die Bereitstellung von gasförmiger Salpetersäure mit hohen Reinheitsgraden bei Badgasdrücken von über 1 bar ermöglicht. Kernstück hierbei ist ein Sättiger, der bei Badgasdrücken von bis zu 100 bar verwendet werden kann.

Der zweite Teil dieser Arbeit befasst sich mit der Kinetik von OH-Reaktionen mit den aliphatischen Ethern Dimethylether (DME), Diethylether (DEE) und Dimethoxymethan (DMM), wobei die Geschwindigkeitskonstanten $k_{\text{DME-d6+OH}}$, $k_{\text{DEE+OH}}$, $k_{\text{DEE-d10+OH}}$ und $k_{\text{DMM+OH}}$ untersucht wurden. Generell wurde bei keiner der gemessenen Geschwindigkeitskonstanten eine Druckabhängigkeit beobachtet. $k_{\text{DME-d6+OH}}$ wurde zwischen 387 und 554 K und 13,0 und 24,4 bar bestimmt. Zur Beschreibung der Temperaturabhängigkeit wurde folgender Ausdruck erhalten: $k_{\text{DME-d6+OH}} = 7,27 \cdot 10^{-23} \left(\frac{T}{\text{K}}\right)^{3,568} \exp\left(\frac{780\,\text{K}}{T}\right)$ cm^3 s^{-1}. Unter den vorliegenden Bedingungen ergibt sich ein Isotopeneffekt von $k_{\text{DME+OH}}/k_{\text{DME-d6+OH}} \approx 3,5$. $k_{\text{DEE+OH}}$ und $k_{\text{DEE-d10+OH}}$ wurden zwischen 295 und 570 K bei 2, 5 und 10 bar untersucht. Die Temperaturabhängigkeit der erhaltenen Werte für $k_{\text{DEE+OH}}$ zeigt eine gute Übereinstimmung mit den direkten Messungen aus der Literatur. Es wurde daher eine Anpassung an all diese Werte durchgeführt, die folgenden Ausdruck ergab: $k_{\text{DEE+OH}} = 1,46 \cdot 10^{-17} \left(\frac{T}{\text{K}}\right)^{1,948} \exp\left(\frac{778\,\text{K}}{T}\right)$ cm^3 s^{-1}. Die erhaltenen $k_{\text{DEE-d10+OH}}$ werden durch folgenden Ausdruck adäquat beschrieben: $k_{\text{DEE-d10+OH}} = 9,82 \cdot 10^{-16} \left(\frac{T}{\text{K}}\right)^{1,334} \exp\left(\frac{343\,\text{K}}{T}\right)$ cm^3 s^{-1}. Der Isotopeneffekt liegt damit bei $k_{\text{DEE+OH}}/k_{\text{DEE-d10+OH}} \approx 2$ im betrachteten Temperaturbereich. Die Messungen der Geschwindigkeitskonstanten $k_{\mathrm{DMM+OH}}$ wurden zwischen 297 und 570 K bei 2, 5 und 10 bar durchgeführt. Die Temperaturabhängigkeit wird durch folgenden Ausdruck wiedergegeben: $k_{\mathrm{DMM+OH}} = 2,90 \cdot 10^{-12} \exp\left(\frac{126\ \mathrm{K}}{T}\right)\ \mathrm{cm}^3\ \mathrm{s}^{-1}$. Bei höheren Temperaturen wurde ein abweichender Verlauf der Daten aus dieser Arbeit und der Literatur beobachtet. Die mechanistischen Implikationen dieser verschiedenen experimentellen Befunde und ihre Konsequenzen für Modellierungen des Zündverhaltens von DMM werden diskutiert.

List of Abbreviations

ϵ .... extinction coefficient
γ .... solid angle of detection
λ .... wavelength
ν .... frequency
ϕ .... quantum yield
σ .... absorption cross section
[S] .... concentration of substrate S
A .... preexponential factor of Arrhenius equation
A_{lk} .... Einstein coefficient of spontaneous emission (transition from l to k)
A/D .... analog-to-digital
Abs .... natural absorbance
ArF .... argon fluoride complex
b .... temperature exponent of modified Arrhenius equation
B_{lk} .... Einstein coefficient of stimulated absorption/emission (transition from l to k)
BBO .... barium borate
BDE .... bond dissociation energy
c .... velocity of light
CC .... coupled cluster
D .... diffusion coefficient
d .... optical path length
DEE .... diethyl ether
DEE-d10 ...... perdeuterated diethyl ether
DFT .... density functional theory
DME .... dimethyl ether
DME-d6 .... perdeuterated dimethyl ether
DMF 2,5-dimethylfuran
DMM .... dimethoxymethane
E_{a} .... activation energy
F .... laser fluence
f .... flow rate

FKM .... fluoroelastomer
fwhm .... full width half maximum
g_i .... degree of degeneracy of state i
GGA .... generalized gradient approximation
h .... Planck constant
H_2O_2 .... hydrogen peroxide
H_2SO_4 .... sulfuric acid
H_p .... primary H atoms
H_s .... secondary H atoms
He .... helium
HF .... Hartree-Fock
HNO_3 .... nitric acid
I .... intensity of electromagnetic radiation
I_0 .... initial intensity of electromagnetic radiation
I_f .... intensity of fluorescence radiation
I_{sat} .... saturation intensity
IC .... internal conversion
ip .... in-plane
ISC .... intersystem crossing
k .... rate coefficient
KNO_3 .... potassium nitrate
KrF .... krypton fluoride complex
LIF .... laser-induced fluorescence
LOD .... limit of detection
MF 2-methylfuran
n .... integer
N_i .... number of molecules in the state i
N_A .... Avogadro constant
NIR .... near infrared
NO_2 .... nitrogen dioxide
NO_x .... collective term for nitrogen oxides
$O(^1D)$ .... oxygen atom in the first excited singulett state
OD .... optical density
OH .... hydroxyl radical
OME .... polyoxymethylene dimethyl ether
op .... out-of-plane
P .... pressure
PLP .... pulsed laser photolysis

PMT .... photomultiplier tube
POMDME .... polyoxymethylene dimethyl ether
PTFE .... polytetrafluoroethylene
Q_{lk} .... Einstein coefficient of quenching (transition from l to k)
R .... universal gas constant
RRKM theory . Rice Ramsperger Kassel Marcus theory
S_i .... singlet state i
SACM .... statistical adiabatic channel model
T .... temperature
t .... time
t-BuOOH ...... *tert*-butyl hydroperoxide
T_i .... triplet state i
TS .... transition state
TST .... transition state theory
UV .... ultraviolet
V .... volume
Vis .... visible
VR .... vibrational relaxation
$\sqrt{x^2}$ .... root mean square distance covered by a diffusing particle
XeCl .... xenon chloride complex
XeF .... xenon fluoride complex
z .... distance

1 Introduction

In atmospheric chemistry hydroxyl radicals (OH) are commonly called the 'detergent of the atmosphere' [1]. In combustion processes one approach to identify the point of ignition is to detect the sudden rise of the OH radical concentration [2]. These two examples illustrate the great importance of these small particles for chemical reactions in the gas phase. But what are OH radicals and where do they come from? What is the reason for their high relevance in gas phase chemistry?

In modern terms, radicals are chemical compounds which exhibit one, or several, unpaired electrons [3]. However, chemical compounds pursue filled electron shells, in which all electrons are paired [4]. Consequently, a general characteristic of radicals is their high reactivity if no stabilizing substituents are present. The OH radical consists of a hydrogen and an oxygen atom connected with a σ-bond. Due to the different electronegativities of oxygen and hydrogen, it exhibits a considerable electric dipole moment of 1.66 D [5]. (In comparison, the electric dipole moment of hydrochloric acid, for example, amounts to only 1.11 D [5].) The unpaired electron is mainly localized at the oxygen atom, while the small hydrogen atom can rarely contribute to a stabilization of the radical. As a result the particle is extremely reactive.

In the gaseous state, chemical reactions proceed typically via radical intermediates. The most prominent representatives of gas phase reactions with high relevance on our daily life can be found in combustion processes or in the atmosphere.

In a combustion process, combustible material with a high energy content reacts with an oxidant, which is commonly the oxygen of the ambient air. Due to the high amount of released energy, it is accompanied with the appearance of flames and sometimes even with the phenomenon of an explosion. From the chemical point of view a combustion process can be described with a radical chain reaction. Thus, it can be divided into initiation, propagation/branching and termination steps. In the chain initiation steps the relatively stable reactants are transformed to reactive radical species by energy input in the form of e.g. an ignition spark or by compression like, for example, in gasoline and diesel engines, respectively. In the propagation and branching steps a formed radical reacts fast with a more stable compound of the reaction mixture (e.g. the fuel molecule itself). The resulting products include one (propagation) or two (branching) other radical species, which further propagate the radical chain. Chain termination steps are radical consuming steps, in which stable products are formed. Only if propagation and especially branching are fast enough to dominate the termination steps and to maintain the radical pool, ignition takes place. [2]

The concentrations of substances with a high content of energy in the atmosphere are too low to enable radical chain reactions which are initiated only by a spatially and temporally limited energy input. Here, the sunlight serves as a continuous energy source at daytime. Some species which are present in the atmosphere are cleaved by the interaction with this electromagnetic radiation forming different kind of radicals. Beside physical effects like transport and deposition, the reaction of trace gases with these radicals are the most important processes which remove e.g. pollutants from the atmosphere. [1]

Hence in both cases, combustion and atmospheric processes, complicated chemical mechanisms possibly consisting of thousands of single reaction steps have to be considered. How can the importance of a single reaction step, like e.g. the reaction of OH with a particular compound, in the whole reaction mechanism be evaluated? Here, chemical reaction kinetics comes into play. It is concerned with the rate of chemical reactions, which are defined by the temporal change of an educt or product concentration. This reaction rate depends on the concentrations of all reacting species. The so-called *rate coefficient* k is the constant of proportionality between the reaction rate and the educt concentrations and is specific for a particular chemical reaction. To assess the influence of e.g. the reaction of a compound with OH on the overall removal of the compound from the atmosphere, the *branching* between competing degradation channels has to be calculated. It quantifies the proportion of the particular reaction on the overall considered process by the ratio of the reaction rates.

What is the detailed role of OH now in a combustion process? Where does the designation 'detergent of the atmosphere' come from?

In combustion processes with oxygen as an oxidant the OH radical is amongst few other radicals the most important species to provoke chain propagation and branching. It can be formed in several initiation sequences, e.g. from the decomposition of intermediate peroxides. The sudden rise of OH radicals therefore indicates that chain propagation and branching dominate and that the ignition is successful. [2, 6]

In atmospheric chemistry OH radicals at daytime are generated mainly from the reaction of excited oxygen ($O(^1D)$) atoms with water. $O(^1D)$ is formed when ozone is cleaved by sunlight. The OH radicals then react with nearly every trace gas which is present in the atmosphere. With this reaction, mostly unpolar educts are transformed into oxidized polar products, which are easily removed from the atmosphere by dissolving in rain droplets or by dry deposition. [1, 7] Many pollutants are eliminated from the atmosphere predominantly by this process. 90 % of the methane removal from the atmosphere, for example, is initiated by OH [8]. After reacting with a trace gas, OH is often regenerated in catalytic cycles. Hence, the concentration of OH in a particular area at daytime is relatively stable. [7] It is therefore often also referred to 'the self-cleansing capacity of the atmosphere' [9].

For a proper description of atmospheric or combustion processes, the detailed reaction mechanism of the particular system is needed. In addition, different physical processes often have to be considered. Complicated differential equation systems result, which have to be solved in elaborate numerical solution procedures. Here, the knowledge of the accurate rate coefficients of all reaction steps involved in the reaction mechanism is of particular importance. In addition, most rate coefficients are dependent on the temperature and some of them also exhibit a pressure dependence. This condition-dependent behavior has to be considered in the models, too.

Due to their significant dipole moment, the OH radical often forms weakly bound complexes with the other reactant molecule before the actual reaction takes place. Such types of reactions are called *complex-forming reactions*. OH radical reactions often exhibit different possible channels for such complex-forming processes, which sometimes additionally compete with direct addition or abstraction reactions. All of these parallel reaction steps possess their own rate coefficient with an individual temperature and pressure dependence. For the entire reaction OH + reactant, complicated correlations of the overall rate coefficient with the temperature and pressure can therefore result.

In modern chemical kinetics different approaches for the determination of rate coefficients and their dependence on the temperature and pressure are known. They can be divided into experimental and theoretical techniques. Typically, in experiments the reaction conditions are varied and the rate coefficient is determined under different conditions. However, the possible range for the variation of e.g. the temperature is limited. As a result, often only parts of the relevant range of conditions can be investigated. Moreover, a prediction of the different reaction channels and their characteristics is difficult solely from experimental results.

Here, the theoretical investigation is an important tool. With quantum chemical calculations a deeper insight into the reaction process and the particular channels can be gained. With the different approaches of statistical rate theory the corresponding rate coefficients can be calculated over a wide range of conditions. However, the theoretical investigation exhibits disadvantages, too. Most calculation steps are not exact due to several approximations and simplifications, which have to be made. Thus, the rate coefficients obtained from theory often only give an indication of the order of magnitude and the rough qualitative dependence on the temperature and pressure. The improvement of the accuracy of the theoretical approaches is still subject of research in chemical kinetics.

As a consequence, a detailed understanding of the kinetics of a particular reaction can often only be gained by the combination of extensive theoretical and experimental studies. In the present work four reactions of OH radicals with different compounds were experimentally investigated.

The first reaction which was studied is the reaction of OH radicals with nitric acid (HNO_3):

$$HNO_3 + OH \xrightarrow{k_1} \text{products.} \tag{R1}$$

Two different aspects determine the relevance of this reaction. On the one hand, it is an important reaction in the atmospheric cycle of nitrogen oxides [10]. On the other hand, it is of practical relevance as nitric acid is a commonly used precursor for OH radicals in pulsed laser photolysis/laser-induced fluorescence experiments [11]. In this case, the reaction of nitric acid with OH is an unavoidable side reaction to the reaction of interest. For both reasons the knowledge of the rate coefficient k_1 is of particular importance. In preliminary studies, it turned out that impurities biased the measurements of k_1 [12] and also some measurements on other systems [13, 14]. As a result, the purity of nitric acid had to be studied first. Important consequences for the general experimental approach were derived, and a device for the supply of pure nitric acid in the gas phase was developed. Moreover, the adequacy of the developed approach was proven in a representative measurement and recommendations for the further experimental procedure were deduced.

The second part of this work was the investigation of the OH reactions with different ethers. Ethers are promising candidates for future biofuels [15, 16]. Moreover, the liquid ethers are widely known as important solvents, which exhibit a risk of violent explosions [17]. The reactions of the compounds dimethyl ether (DME), diethyl ether (DEE) and dimethoxymethane (DMM) with OH radicals were studied in this work:

$$\text{DME}(-d_6) + OH \xrightarrow{k_{2(d)}} \text{products} \tag{R2}$$

$$\text{DEE}(-d_{10}) + OH \xrightarrow{k_{3(d)}} \text{products} \tag{R3}$$

$$\text{DMM} + OH \xrightarrow{k_4} \text{products.} \tag{R4}$$

The focus of these studies was put on different aspects. On the one hand, theoretical methods which were applied in the calculations of Kiecherer [18] were validated by the investigation of the isotope effect of the rate coefficients k_2 and k_3, and a deeper insight in the general reaction characteristics of ethers with OH radicals was gained. Moreover, especially the rate coefficients k_3 and k_4 are only known in narrow pressure and temperature ranges. Here, the aim was to expand the range of conditions in which these rate coefficients are determined experimentally, focusing on ignition-relevant conditions.

2 Fundamentals

2.1 Linear and Non-Linear Spectroscopy

For the spectroscopic methods applied in this work, linear and non-linear effects during the interaction between matter and electromagnetic radiation in the ultraviolet (UV) and visible (Vis) region of the spectrum were important. The general fundamentals of these different types of spectroscopy are explained in this section. The detailed discussion is limited to a simplified model of a two-level system, as all basic characteristics can be illustrated most clearly with this. The relevant deviations and limits are discussed briefly. For a more detailed approach, the reader is referred to the literature (see e.g. refs. [19, 20]).

2.1.1 Excitation and Following Processes

The absorption of UV/Vis radiation leads to the excitation of electrons into higher electronic states. Thereupon, different relaxation processes can follow, which are illustrated schematically in the Jablonski diagram in figure 2.1.

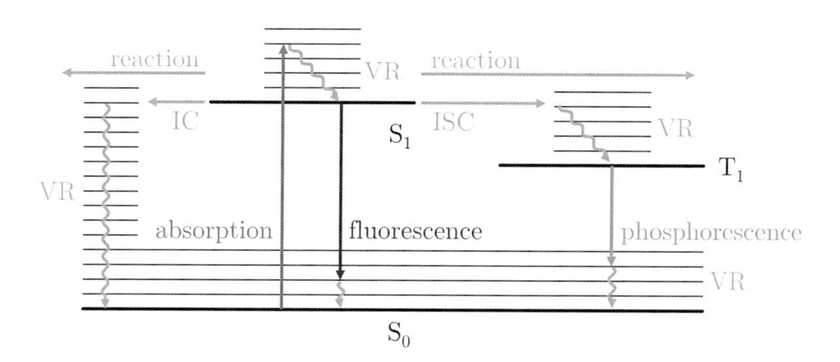

Figure 2.1: Schematic Jablonski diagram illustrating typical processes which follow an electronic excitation; IC: internal conversion, ISC: intersystem crossing, VR: vibrational relaxation.

Usually, the ground state of a molecule is a singlet state (S_0). Nevertheless, there are also examples of molecules which exhibit another spin multiplicity of the electronic ground state. In any case, according to the selection rules the excitation occurs into a higher electronic state with the same spin multiplicity (S_1 in figure 2.1) and often also in excited vibrational (and rotational) states.

As stated by the rule of Kasha [21], the vibrational (and rotational) relaxation is very fast, so that all internal processes connected with the emission of radiation take place from the vibrational ground state of the excited electronic state. When a molecule fluoresces subsequently the excited electron falls back into the electronic ground state under spontaneous emission. Stimulated emission is an analogous process, whereas the relaxation is not spontaneously but induced by radiation. Another relaxation process under emission of radiation is phosphorescence. It is preceded by an intersystem crossing into an electronic state of another spin multiplicity (triplet state T_1 in figure 2.1), which is formally spin-forbidden. The subsequent transition into the electronic ground state is spin-forbidden as well, which is why phosphorescence lifetimes are typically significantly higher than fluorescence lifetimes.

In addition, under perpetuation of the spin multiplicity the excited molecule can undergo an internal conversion, in which the excited electron passes over to a highly excited vibrational (and rotational) state of the electronic ground state. Subsequently, it relaxes radiationless to the vibrational ground state. This process is induced by collisions with other particles and can be a competitive process to fluorescence. In this case, it is referred to as dynamic fluorescence quenching.

After the absorption of a photon, the molecule can also react in chemical reactions. The reaction can take place directly after excitation or can be preceded by other processes like e.g. intersystem crossing.

2.1.2 Excited State Population in a Two-Level System

The basic relations regarding the population of the excited state by absorption can be best illustrated with a two-level system. The discussion is mainly based on refs. [20, 22, 23].

In general, the transition probabilities of the excitation and relaxation processes are quantified by Einstein coefficients. The Einstein coefficient of the absorption is given by B_{kl}. In the idealized two-level system three different relaxation processes compete with each other after excitation. On the one hand, stimulated emission (B_{lk}) can occur. Its relaxation rate as well as the excitation rate of the absorption are dependent on the energy density of the laser, which is the laser intensity I divided by the velocity of light c. On the other hand, two processes exist, whose rates are independent of the laser intensity. These are the spontaneous emission or fluorescence (A_{lk}) and the radiationless quenching induced by collisions (Q_{lk}). The processes and their Einstein coefficients are illustrated in Figure 2.2.

Due to the generally high energy gap between electronic ground and excited states, all molecules are assumed to be in the ground state before the interaction with electromagnetic radiation.

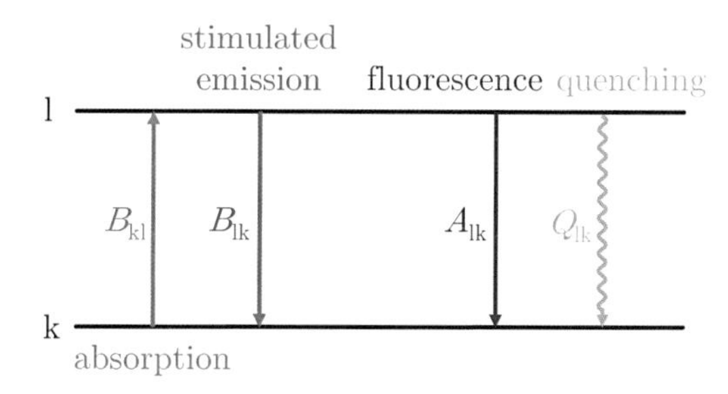

Figure 2.2: Two-level model with illustration of the excitation and the different relaxation processes and their corresponding Einstein coefficients.

Hence, the temporal change of the number of excited molecules N_{l} is given by

$$\frac{\mathrm{d}N_{\mathrm{l}}}{\mathrm{d}t} = B_{\mathrm{kl}}\frac{I}{\mathrm{c}}N_{\mathrm{k}} - \left(B_{\mathrm{lk}}\frac{I}{\mathrm{c}} + A_{\mathrm{lk}} + Q_{\mathrm{lk}}\right)N_{\mathrm{l}}\,, \tag{2.1}$$

where N_{k} is the number of molecules in the ground state.

After a short introduction time a steady-state between excitation and relaxation can be assumed:

$$\frac{\mathrm{d}N_{\mathrm{l}}}{\mathrm{d}t} = 0\,. \tag{2.2}$$

In combination with the balance equation

$$N_{\mathrm{tot}} = N_{\mathrm{k}} + N_{\mathrm{l}}\,, \tag{2.3}$$

with N_{tot} being the total number of molecules in the system, the following expression results:

$$N_{\mathrm{l}} = \frac{B_{\mathrm{kl}}\frac{I}{\mathrm{c}}}{(B_{\mathrm{kl}} + B_{\mathrm{lk}})\frac{I}{\mathrm{c}} + A_{\mathrm{lk}} + Q_{\mathrm{lk}}}N_{\mathrm{tot}}\,. \tag{2.4}$$

By rearranging equation 2.4 and defining a so-called saturation intensity I_{sat} [23]

$$I_{\mathrm{sat}} = \frac{(A_{\mathrm{lk}} + Q_{\mathrm{lk}})\,\mathrm{c}}{B_{\mathrm{kl}} + B_{\mathrm{lk}}}\,, \tag{2.5}$$

the simplified expression

$$N_{\mathrm{l}} = \frac{B_{\mathrm{kl}}}{B_{\mathrm{kl}} + B_{\mathrm{lk}}}\frac{1}{1 + \frac{I_{\mathrm{sat}}}{I}}N_{\mathrm{tot}} \tag{2.6}$$

is obtained.

The relation between B_{kl} and B_{lk} is given by:

$$\frac{B_{\mathrm{kl}}}{B_{\mathrm{lk}}} = \frac{g_{\mathrm{k}}}{g_{\mathrm{l}}}\,, \tag{2.7}$$

where g_i is the degree of degeneracy of the corresponding electronic state. In a simple two-level system g_{i} is obviously 1 for both levels. Thus, B_{kl} equals B_{lk} and equation 2.6 becomes

$$N_{\mathrm{l}} = \frac{1}{2}\frac{1}{1+\frac{I_{\mathrm{sat}}}{I}}N_{\mathrm{tot}}\,. \tag{2.8}$$

At this point, different cases can be distinguished. If I is very low ($I << I_{\mathrm{sat}}$), N_{l} approaches zero. In this case, the perturbation of the system by the absorption of light is negligible and, in good approximation, all molecules are in the ground state. The emitted intensities from conventional non-coherent light sources are usually so low that the absorption processes can be described with this limiting case. The corresponding spectroscopic methods are assigned to *linear* spectroscopy, as the measuring signal changes linearly with the variation of the irradiated light intensity.

Intermediate and very high intensities can usually only be achieved by coherent light sources, i.e. lasers. If I is in the order of magnitude of I_{sat}, then N_{l} is dependent on the light intensity. In contrast, if I is very high ($I >> I_{\mathrm{sat}}$), the whole intensity-containing term in equation 2.8 approaches 1. Thus, N_{l} becomes independent of I and an equal distribution between ground and excited state is approached in the simple two-level system. In these two cases the measuring signal of an spectroscopic experiment does not change linearly with the initial light intensity. More complex correlations, or an independence of the measuring signal from the irradiated intensity, results. Hence, in these cases it is referred to as *non-linear* spectroscopy with the further specification of *saturated* spectroscopy when the limiting case of very high intensities holds.

2.1.3 Linear Absorption Spectroscopy

In absorption spectroscopy, the initial intensity I_0 and the attenuated intensity I is measured when light passes through an absorbing substrate S. The attenuation of the intensity dI in the distance dz is given by [20]:

$$\mathrm{d}I = \sigma I\left(\frac{N_{\mathrm{k}}}{V} - \frac{g_{\mathrm{k}}}{g_{\mathrm{l}}}\frac{N_{\mathrm{l}}}{V}\right)\mathrm{d}z\,. \tag{2.9}$$

Here, V is the illuminated volume and σ is the absorption cross section, which is characteristic for a chemical compound and reflects its absorption capacity at a given wavelength.

For the limiting case of very high light intensities in a two-level system, as described above, the sample becomes transparent, as $N_{\mathrm{l}} = N_{\mathrm{k}}$, and no light attenuation is detected anymore.

In the case of very low intensities, N_l can be neglected. As a result, $\frac{N_\mathrm{k}}{V}$ equals the macroscopic substrate concentration [S]. Integrating over the optical path length d, the following expression results:

$$-\ln\left(\frac{I}{I_0}\right) = \mathrm{Abs} = \sigma[\mathrm{S}]d\,. \tag{2.10}$$

This is the famous Beer-Lambert law, which has been discovered empirically early in the history of physics in the works of Bouguer [24], Lambert [25] and Beer [26]. The negative natural logarithm of the ratio between the leaving intensity I and the entering intensity I_0 is called the natural absorbance, Abs. The equation shows that the detected intensity behind the sample changes linearly with the initial intensity. It enables the determination of the substrate concentration by absorption measurements with non-coherent light sources.

Often, the law refers to the decadic absorbance, which is also called the optical density, OD. The conversion factor from the natural to the decadic logarithm is considered in the constant of proportionality, which is referred to as the extinction coefficient ϵ in the decadic case. It should be kept in mind that, in gas phase spectroscopy, the absorption cross section is commonly used for the characterization of the absorptivity of gaseous compounds [27]. However, conventional spectrometers usually give out the decadic absorbance.

2.1.4 Saturated Laser-Induced Fluorescence

In fluorescence spectroscopy a substrate is studied by selective excitation and detection of the emitted fluorescence. The intensity of the fluorescence I_f is given by [22]:

$$I_\mathrm{f} = A_\mathrm{lk}\mathrm{h}\nu\frac{\gamma}{4\pi}VN_\mathrm{l}\,. \tag{2.11}$$

Here, h is the Planck constant, ν the frequency of the light source and γ the solid angle of the collection optics. It shows that I_f is proportional to the population N_l of the excited state.

If lasers are applied as a light source, generally a non-linear behavior is expected. According to the discussion of the two-level system in subsection 2.1.2, N_l is dependent on the laser output at lower intensities. At higher intensities, saturation conditions are achieved. In this case I_f becomes independent of I and its instabilities, as well as of quenching processes. A significantly improved signal-to-noise ratio results in comparison to non-saturated LIF.

However, the model of a two-level system is a simplification, which cannot describe the situation in a real molecule exactly, especially at high pressures. In general, vibrational and rotational transitions have to be considered too [23]. Moreover, especially under high-pressure conditions, pressure broadening of the fluorescence band cannot be neglected. [23, 28]

In addition, a homogeneous distribution of the laser intensity is not realistic on both spatial and temporal scales if a pulsed laser is applied. Furthermore, the spectral resolution of the excitation laser beam is finite. All factors lead to the appearance of so-called 'wings' which characterize the boundary areas in all of the named dimensions. In these 'wings', the intensity is too low for saturation. Thus, there are always parts of the laser pulse in which saturation conditions do not hold. [23]

On the whole, a complete independence of the fluorescence intensity from laser intensity fluctuations and quenching processes cannot be achieved. This is problematic, particularly for quantitative measurements of the OH concentration. However, a quantitative determination of [OH] was not necessary for this work. For relative measurements, like in the present work, the knowledge of the fundamentals of saturated LIF is useful to achieve an improved signal-to-noise ratio by maximizing the laser output.

Nevertheless, especially for high-pressure measurements, fluorescence quenching is still an important issue. On the one hand, Q increases with the pressure. Thus, higher laser intensities are needed to achieve saturation conditions. Additionally, the remaining dependency on quenching in the wings influences the fluorescence intensity, even if saturation is achieved in the central region of the laser pulse. Hence, a reasonable signal strength can only be achieved by the choice of an ineffective quencher, like helium, as bath gas. Moreover, when the radical precursor is selected its quenching capacity has to be borne in mind too. Significantly differing signal-to-noise ratios of the LIF signals can be obtained for different precursors under otherwise the same conditions.

2.2 Temperature and Pressure Dependence of Rate Coefficients of Gas Phase Reactions

The knowledge of the dependence of rate coefficients on the temperature and pressure is of fundamental importance for the understanding of chemical processes in the gas phase. In the last century, the theoretical description of these correlations has been subject of intensive research. Different theories have been developed and improved, so that a detailed understanding of the theoretical background of reaction kinetics could be gained. As the studies in this work were merely of experimental nature, mainly the phenomenological context is discussed here. The relevant approaches for a theoretical description are only mentioned briefly.

2.2.1 Temperature Dependence

It was first found empirically by Arrhenius [29] that the temperature dependence of the rate coefficient $k(T)$ obeys the following law, which is nowadays referred to as the *Arrhenius equation*:

$$k(T) = A \exp\left(-\frac{E_{\mathrm{a}}}{\mathrm{R}T}\right) . \tag{2.12}$$

Here, R is the universal gas constant and T represents the temperature. The parameter A is called the preexponential factor and E_{a} the activation energy. It was assumed by Arrhenius that these were independent of temperature.

The Arrhenius equation can be interpreted roughly by considering a simple bimolecular elementary step. Here, a reaction takes place when the two reactant molecules collide and contain enough energy to overcome the reaction barrier. In the Arrhenius equation the preexponential factor A can therefore be interpreted as a collision parameter, which is proportional to the collision frequency. The exponential term reflects the probability that the energy of a collision is high enough to lead to a reaction. [30]

However, many examples were found which deviate from the simple Arrhenius behavior. Most easily, this becomes obvious when the logarithm of the rate coefficient is plotted against the inverse temperature. In such *Arrhenius plots* a linearity between the plotted values can be observed for the simple Arrhenius behavior. In the case of a deviation, a curved behavior or in some special cases even more complicated correlations can be observed. Often, the temperature dependence in such deviating cases can be described with the *modified Arrhenius equation* [31]:

$$k(T) = A \left(\frac{T}{\mathrm{K}}\right)^{b} \exp\left(-\frac{E_{\mathrm{a}}}{\mathrm{R}T}\right) . \tag{2.13}$$

Here, a third parameter b is introduced and the preexponential term becomes temperature dependent as well.

In the case of simple bimolecular elementary steps the deviation can be explained mostly with the transition state theory (TST) and quantum chemical effects, like tunneling. However, a considered reaction often does not consist of a single elementary step if a non-Arrhenius behavior is observed. Many reactions in the gas phase actually exhibit several consecutive and/or parallel reaction steps. One example are complex-forming reactions, which are described in subsection 2.3. In these cases the Arrhenius equation and its modification have to be treated as an empirical parametrization approach. A physical interpretation of the parameters is often not reasonable and a theoretical understanding of the reaction dynamics and kinetics can only be gained from statistical rate theory. A discussion of the temperature-dependence of rate coefficients from the perspective of modern chemical kinetics and its connection with the historical approach of the Arrhenius equation is given in ref. [32].

2.2.2 Pressure Dependence

Rate coefficients exhibit a pressure dependence if a collision with any collision partner is needed to transfer energy during the reaction process. The simplest example of a pressure dependent reaction is a unimolecular elementary reaction, e.g. the decomposition of a compound. Here, the energy which is needed for the bond cleavage has to be transferred to the molecule before the reaction can take place. If no other activation processes, like the absorption of light, are possible, the energy transfer can only be achieved via collisions. In this case, it is also referred to as *thermally activated* unimolecular reactions.

Lindemann [33] first proposed a general theory to describe this collisional activation process in unimolecular reactions. In a first step, the molecule X is energized by a collision with an arbitrary collision partner M. The activated molecule X* can then be deactivated again by another collision or react in the actual unimolecular reaction forming the product P:

$$\mathrm{X} + \mathrm{M} \underset{k_{\text{-a}}}{\overset{k_{\mathrm{a}}}{\rightleftharpoons}} \mathrm{X}^{*} + \mathrm{M} \tag{R5}$$

$$\mathrm{X} \xrightarrow{k_{\mathrm{b}}} \mathrm{P}. \tag{R6}$$

Assuming a steady-state for X*, the rate of the product formation is:

$$\frac{\mathrm{d[P]}}{\mathrm{d}t} = k_{\mathrm{uni}}[\mathrm{X}] = \frac{k_{\mathrm{a}}k_{\mathrm{b}}}{k_{\text{-a}} + \frac{k_{\mathrm{b}}}{[\mathrm{M}]}}[\mathrm{X}]\,. \tag{2.14}$$

At the limiting case of very high pressures the observed unimolecular rate coefficient becomes:

$$k_{\mathrm{uni}}^{\infty} = \frac{k_{\mathrm{a}}k_{\mathrm{b}}}{k_{\text{-a}}}\,. \tag{2.15}$$

It is therefore independent of the pressure and the overall reaction is of first order. At very low pressures k_{uni} becomes

$$k_{\mathrm{uni}}^{0} = k_{\mathrm{a}}[\mathrm{M}]\,. \tag{2.16}$$

Thus, it changes linearly with the pressure and the reaction order is two in this case. The intermediate region between the high- and low-pressure limit is called the fall-off regime.

For a more precise description of the kinetics of such reactions, the dependence of the rate coefficients on the internal energy of X has to be considered. Due to this dependence, a parametrization of an experimentally determined fall-off curve according to equation 2.16 does not suffice. Different approaches to parametrize the pressure dependence of rate coefficients were developed, e.g. the Troe formalism [34]. The calculation of pressure dependent rate coefficients is possible on the basis of statistical rate theory.

2.2.3 Statistical Rate Theory

In general, statistical rate theory provides a statistical mechanical approach for the calculation of rate coefficients under various conditions on the basis of a known potential energy surface (or at least its important stationary points) of the considered reaction. Different approaches for different types of chemical reactions have been developed.

For the description of the kinetics of pressure independent reactions, or of reactions at the high-pressure limit, canonical transition state theory (TST) [35, 36] can be applied. Here, the reactants and intermediate species can be assumed to exhibit a Boltzmann distribution.

If a rate coefficient is pressure dependent, a thermal equilibrium cannot be assumed anymore for all species which are involved in the chemical reaction. Thus, it has to be changed from a canonical description of the reaction process to a microcanonical approach. Here, a dependence of the rate coefficient on the internal energy of the reacting compounds has to be assumed. In this case each particular internal energy implicates a different microscopic rate coefficient. Different approaches exist for the description of such microscopic rate coefficients, in which it has to be distinguished between so-called tight and loose transition states. If the reaction coordinate exhibits a maximum of potential energy, a defined transition state structure can be located. It is called a tight transition state. In contrast, it is referred to as loose transition state if no maximum is existent along the reaction coordinate. For reactions with a tight transition state the Rice-Ramsperger-Kassel-Marcus (RRKM) theory [37–39] is the approach most often used. If reactions with a loose transition state are considered, the microscopic rate coefficient can be described, for example, with the statistical adiabatic channel model (SACM) [40].

To obtain a macroscopic rate coefficient from these microcanonical values, the system is formally divided into discrete energy levels, as especially the number of actual energy levels of the rotation is usually too high to be considered individually. The temporal evolution of the population of each energy level is then considered. The ensemble of these differential equations is called the master equation [41, 42]. The solution of this equation system gives the macroscopic rate coefficient.

A detailed description of the different theoretical approaches is beyond the scope of this work. For the fundamentals of statistical rate theory see e.g. refs. [43–45].

2.3 Complex-Forming Bimolecular Reactions

Complex-forming bimolecular reactions are a frequently appearing reaction type in gas phase chemistry. In this section the characteristics of their kinetics are discussed qualitatively and the, sometimes curious, temperature and pressure dependencies of the corresponding rate coefficients which are experimentally observable are explained. For a description of the theoretical background, the reader is referred to the literature (see e.g. refs. [43, 46–48]).

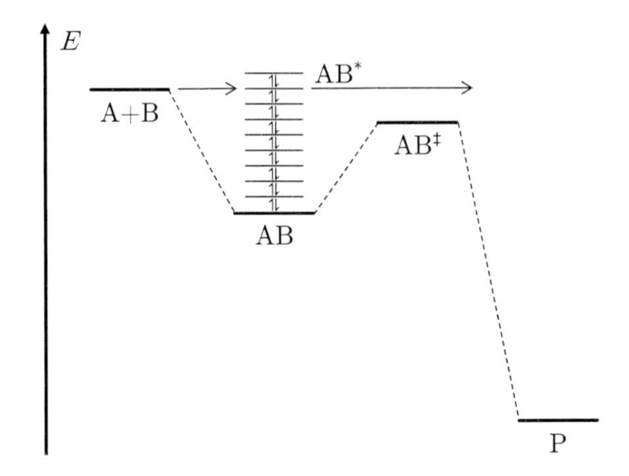

Figure 2.3: Schematic potential energy diagram of a complex-forming bimolecular reaction. The arrows illustrate the reaction and stabilization processes.

2.3.1 Mechanism and Overall Kinetics

Complex-forming bimolecular reactions can be described with the following basic reaction scheme:

$$A + B \underset{k_{\text{-a}}}{\overset{k_{\text{a}}}{\rightleftharpoons}} AB^* \tag{R7}$$

$$AB^* \underset{k_{\text{act}}}{\overset{k_{\text{stab}}}{\rightleftharpoons}} AB \tag{R8}$$

$$AB^* \xrightarrow{k_{\text{b}}} P. \tag{R9}$$

Thus, in a first reaction step an energized complex AB^* is formed from the educts A and B. AB^* can then be stabilized to AB, which can be re-activated again. Moreover, the excited complex can react in an irreversible reaction to the product(s) P. Such type of reactions are also referred to as *chemically activated* unimolecular reactions, because the energy for the second unimolecular reaction step AB→P is provided by the initial chemical reaction step.

A schematic potential energy diagram of a simple complex-forming bimolecular reaction is shown in figure 2.3. It illustrates the competition between the forward reaction of the energized complex and its stabilization. The first reaction step is often a simple barrierless recombination reaction, while the second reaction step exhibits a reaction barrier and a tight transition state.

The reaction rates of the educt and intermediate species are given by

$$\frac{\mathrm{d[A]}}{\mathrm{d}t} = -k_{\mathrm{a}}[\mathrm{A}][\mathrm{B}] + k_{\text{-a}}[\mathrm{AB}^*] \tag{2.17}$$

$$\frac{\mathrm{d[AB^*]}}{\mathrm{d}t} = k_{\mathrm{a}}[\mathrm{A}][\mathrm{B}] - (k_{\text{-a}} + k_{\mathrm{stab}} + k_{\mathrm{b}})\,[\mathrm{AB}^*] + k_{\mathrm{act}}[\mathrm{AB}] \tag{2.18}$$

$$\frac{\mathrm{d[AB]}}{\mathrm{d}t} = k_{\mathrm{stab}}[\mathrm{AB}^*] - k_{\mathrm{act}}[\mathrm{AB}]\,. \tag{2.19}$$

Assuming that only educts are present when the reaction starts, it can be differentiated between different steady-states at different reaction times. After a short induction period, the so-called *intermediate steady-state* is valid. It is followed by the *final steady-state* at later reaction times. [46]

In the intermediate steady-state, the steady-state assumption only holds for $[\mathrm{AB}^*]$. Moreover, due to the lack of significant amounts of the stabilized complex AB, the rate of the re-activation reaction $k_{\mathrm{act}}[\mathrm{AB}]$ is negligible. Thus, the steady-state concentration of AB^* is given by the expression

$$[\mathrm{AB}^*] = \frac{k_{\mathrm{a}}}{k_{\text{-a}} + k_{\mathrm{stab}} + k_{\mathrm{b}}}[\mathrm{A}][\mathrm{B}]\,. \tag{2.20}$$

Hence, the rate coefficient k_{obs} which is observable by means of the educt concentration decrease in the intermediate steady-state is described by

$$k_{\mathrm{obs}} = \frac{k_{\mathrm{b}} + k_{\mathrm{stab}}}{k_{\text{-a}} + k_{\mathrm{b}} + k_{\mathrm{stab}}} k_{\mathrm{a}}\,. \tag{2.21}$$

It can be shown that the transition from the intermediate to the final steady-state takes place at a reaction time which corresponds roughly to the mean lifetime of AB. This mean lifetime is the inverse overall thermal rate coefficient for the unimolecular decomposition of the complex. [46]

In the final steady-state the steady-state assumption is valid for both $[\mathrm{AB}^*]$ and $[\mathrm{AB}]$. Thus, $k_{\mathrm{stab}}[\mathrm{AB}^*]$ equals $k_{\mathrm{act}}[\mathrm{AB}]$ and the steady-state concentration of AB^* is given by

$$[\mathrm{AB}^*] = \frac{k_{\mathrm{a}}}{k_{\text{-a}} + k_{\mathrm{b}}}[\mathrm{A}][\mathrm{B}]\,. \tag{2.22}$$

For the observable rate coefficient in the final steady-state the following expression results:

$$k_{\mathrm{obs}} = \frac{k_{\mathrm{a}} k_{\mathrm{b}}}{k_{\text{-a}} + k_{\mathrm{b}}}\,. \tag{2.23}$$

The model described above is a simplification, which only includes the existence of a energetically activated species AB^* and an equilibrated species AB. However, analogously to a thermally activated unimolecular reaction, different energetic levels of AB have to be considered for a precise description. These are indicated in figure 2.3. The rate coefficients of the unimolecular processes have to be treated dependent on the internal energy of AB/AB*, and the temporal evolution of the population of the different energy levels has to be considered in master equation calculations. Here, a time-independent approach can be formulated for the intermediate and final steady-state. If none of the limiting cases are valid, the time-dependent master equation has to be solved. [46]

2.3.2 Pressure Dependence

To understand the pressure dependence of the kinetics of complex-forming bimolecular reactions, the different limiting cases have to be accounted for separately.

In the intermediate steady-state the pressure dependence is dominated by the stabilization process. Thus, the pressure dependence of the observable rate coefficient can be explained qualitatively with the simplified model mechanism described above under the assumption of k_{stab} being the only pressure-dependent rate coefficient. If the pressure is very low ($k_{\mathrm{stab}}(P,T) << k_{\text{-a}}(T)+k_{\mathrm{b}}(T)$), the stabilization can be neglected and only the reactive processes contribute to the observable rate coefficient, yielding

$$k^{0}_{\mathrm{obs}}(T) = \frac{k_{\mathrm{b}}(T)}{k_{\text{-a}}(T)+k_{\mathrm{b}}(T)} k_{\mathrm{a}}(T)\,. \tag{2.24}$$

In the case of high pressures ($k_{\mathrm{stab}}(P,T) >> k_{\text{-a}}(T)+k_{\mathrm{b}}(T)$), k_{stab} dominates over the rate coefficients for the chemical reactions. Consequently, it cancels out in the expression for the observable rate coefficient, which simplifies to

$$k^{\infty}_{\mathrm{obs}}(T) = k_{\mathrm{a}}(T)\,. \tag{2.25}$$

As a result, k_{obs} is independent of the pressure for both the high- and the low-pressure limiting cases. Because rate coefficients are only defined for positive values, $k_{\mathrm{b}}/(k_{\text{-a}}+k_{\mathrm{b}}) < 1$ applies. Hence, the low-pressure limit is always lower than the high-pressure limit in the intermediate steady-state. Analogous to thermally activated reactions, the regime in between is called the fall-off regime. Altogether, an s-shaped curve for the pressure dependence of the observable rate coefficient of complex-forming bimolecular reactions in the intermediate steady-state results.

Strictly speaking, the rate coefficients $k_{\text{-a}}$ and k_{b} of the unimolecular reactive processes exhibit a pressure dependence, too. It results from their dependence on the energy distribution of the intermediate complex AB. Again, two different limiting cases exist. At very low pressures no stabilization occurs and the energy distribution of AB becomes independent of the pressure. At high pressures the energy distribution of AB approaches a thermal equilibrium. Hence, in both of these limiting cases $k_{\text{-a}}$ and k_{b} are independent of the pressure. In between, a fall-off behavior can be observed. However, this pressure dependence is small compared to the one of the stabilization process and can thus be neglected in the intermediate steady-state.

In the final steady-state there is no more net stabilization and k_{obs} becomes independent of k_{stab}. In consideration of the pressure dependence of $k_{\text{-a}}$ and k_{b}, equation 2.23 can be rewritten as:

$$k_{\mathrm{obs}}(P,T) = k_{\mathrm{a}}(T)\frac{1}{1+\frac{k_{\text{-a}}(P,T)}{k_{\mathrm{b}}(P,T)}}\,. \tag{2.26}$$

This expression shows that here a general prediction of the positions of the high- and low-pressure limits relative to each other cannot be made. If $k_{\text{-a}}(P,T)$ exhibits a weaker pressure dependence than $k_{\mathrm{b}}(P,T)$, the low-pressure limit is lower than the high-pressure limit. In the other case, the low-pressure limiting rate coefficient is situated above the high-pressure limit. Thus, an s-shaped or reversed s-shaped curve (with a negative pressure dependence in the fall-off regime) for the pressure dependence of k_{obs} can occur for complex-forming bimolecular reactions in the final steady-state.

2.3.3 Temperature Dependence

The temperature dependence of complex-forming bimolecular reactions is mainly governed by the barrier heights of the individual reaction steps. If the first association step (k_{a}) is assumed to be barrierless, like it is the case in figure 2.3, its rate coefficient is roughly independent of the temperature. The reverse ($k_{\text{-a}}$) and forward (k_{b}) reactions of the prereactive complex AB both exhibit a reaction barrier and therefore a positive temperature dependence of their rate coefficients. The extent of this temperature dependence correlates with the barrier height. The higher the energy barrier the more distinct the increase of the rate coefficient with the temperature.

However, the reverse and forward reaction of the prereactive complex AB are competitive reaction steps, which lead to the educts in the first case and the products in the latter. Thus, for the temperature dependence of the observable rate coefficient k_{obs}, two different cases have to be distinguished. If the forward barrier is lower than the reverse barrier, the temperature dependence of the reverse reaction is more pronounced. Hence, in the case of a temperature increase, the re-formation of the educts is more accelerated than the formation of the products. As a result, the observable rate coefficient k_{obs} exhibits a negative temperature dependence. In the case of a higher barrier of the forward reaction compared to the reverse reaction, the acceleration of the product-formation with increasing temperature is more pronounced and k_{obs} is positively dependent on the temperature.

Of course, the barrier heights are not the only influence contributing to the temperature dependence of the observable rate coefficients. Again, a precise description can only be achieved by master equation calculations. However, a rough estimation of the relative position of the energy barriers in complex forming bimolecular reactions from e.g. experimentally determined rate coefficients is feasible.

Care has to be taken when different reaction channels compete with each other in a studied reaction. In this case, the rate coefficients of all relevant channels contribute to the observed rate coefficient. This superposition can lead to complicated correlations of the observable rate coefficient with the temperature, which might deviate significantly from a classical Arrhenius behavior. Here, an interpretation of the temperature dependence of k_{obs} with respect to the characteristics of the potential energy diagram is not reasonable if no other information about the possible reaction channels is available.

2.3.4 Quantum chemical Methods for the Description of Potential Energy Surfaces

A precise knowledge of the potential energy surface of a complex-forming bimolecular reaction is the basis for a reliable theoretical description of its kinetics. Thus, the quantum chemical method for the calculation has to be selected carefully.

In principle, the methods which are typically used for this purpose can be divided into (post-)Hartree-Fock (HF) and density functional theory (DFT) methods. The basic characteristics of these general approaches, and the concrete methods which are relevant for this work, are briefly described in this subsection. For detailed explanations and an extensive overview of different methods the reader is referred to ref. [49].

The HF approach is based on the solution of the time-independent Schrödinger equation [50]. Here, the application of the Born-Oppenheimer approximation [51] is a fundamental issue for chemical systems. According to this, the light electrons move much faster than the heavy nuclei so that their motion can be treated separately. Thus, on the time scale of an electron motion the nuclei are assumed to be spatially fixed and the electronic Schrödinger equation can be formulated for constant coordinates of the positions of the nuclei.

Nevertheless, the description of the dynamics of several electrons, which all interact with each other, is still an elaborate treatment. In the basic HF approach [52] the Coulomb interactions are therefore only considered with an average potential. Apart from that, the kinetic energy of a single electron is treated independently of electron-electron interactions. However, with the use of Slater determinants as wave functions, the electron exchange interactions are considered exactly.

The use of average electron-electron interactions is the most problematic drawback of the HF method. For this reason several methods were developed which account for the electron correlation more accurately. One popular approach is a more extensive description of the wave functions. The methods based on this principle are called electron correlation methods or post-HF methods. [49]

One post-HF method is the coupled cluster (CC) approach [53]. Here, a reference wave function, based on a Slater determinant, is expanded with an exponential operator, which is composed of a sum of operators for the different levels of excitation. This sum is truncated after the nth excitation operator, where n defines the level of theory. In the case of a truncation after the operator for the second excitation for example, the method is referred to as CCSD, where S denotes the single and D the double excitation. With every higher excitation operator which is considered the calculation becomes more accurate but also much more costly.

In the density functional theory the electron density is of fundamental importance. It is based on the Hohenberg-Kohn theorem [54], which states that the electron density suffices for a complete determination of the ground state electronic energy. This exhibits the great advantage that, in comparison to wave functions, the number of unknown variables is significantly reduced.

In the first attempts to develop calculation methods on the basis of this theorem, all energy terms were described as a functional of the electron density. However, the description of the kinetic energy of the electrons is poor in these approaches. As a result, the early DFT methods always stayed behind post-HF methods in the field of chemistry and thus were not applied practically for chemical problems. [49]

The breakthrough for DFT methods was achieved with a suggestion by Kohn and Sham [55], that the kinetic energy of the electrons can be described with the concept of molecular orbitals instead of the pure electron density. With this approach, the most prominent problems of the orbital-free models could be overcome. The disadvantage is that the number of unknown variables increases and therefore the effort of calculation. Nevertheless, in many cases the modern DFT methods deliver good results with low computational cost compared to post-HF methods.

A problem which still remains is that no strictly physical approach for the description of the electron-correlation energy as a functional of the electron density exists. Thus, this energy term has to be approximated somehow. The methods which describe this functional on the basis of purely theoretical considerations do not give satisfactory results so far. The more successful approach is the approximation of the electron-correlation energy with the help of empirical parameters. Here, relatively simple approaches perform remarkably well. As a result, the DFT methods for practical application are not *ab initio* methods, but contain experimentally obtained information. Hence, a general problematic matter of DFT is still the lack of a systematic approach for the improvement of the results towards an exact solution. [56]

Commonly used types of DFT functionals are so-called hybrid functionals [57]. Here, the exchange-correlation energy is described by a linear combination of HF and DFT terms in order to exploit error compensation. The B3LYP functional [58], for example, contains terms based on HF theory as well as on the generalized gradient approximation (GGA). The contribution of the different terms is weighted by empirical factors.

For the description of potential energy surfaces of complex-forming bimolecular reactions, the coupled cluster approach gives very stable results provided that a multireference character of the wave function can be excluded. However, these methods are very costly and their application in systems with a high number of atoms can be infeasible. [59] If a cheaper method for the calculation of the potential energy surface is required, the B3LYP functional is often applied. In many cases, this popular DFT functional performs well. However, dispersion interactions generally are only poorly described with B3LYP [60]. Also, the computation of transition state energies and structures with the B3LYP functional turned out to be flawed in many cases [61–63]. Hence, care has to be taken when B3LYP is used for the description of weakly-bound complexes and their reaction kinetics. It is strongly recommended to test the performance of the B3LYP functional for the description of a particular type of reaction before it is applied for an uncharted system [56].

3 Experimental

For spectroscopic experiments in the gas phase under slow-flow conditions, the different gaseous compounds of interest have to be injected into the measuring cell in defined concentrations and with a controlled flow rate. For this purpose, different techniques were applied, which are explained at the beginning of this chapter.

All rate coefficients determined in the present work were measured with pulsed laser photolysis/laser-induced fluorescence (PLP/LIF) technique at pressures above 1 bar. The experimental setup which was used is also presented in this chapter.

Preliminary measurements regarding the purity of the OH precursor nitric acid were necessary. Therefore, a new experimental setup for the detection of this compound and its impurities was built up applying an absorption spectrometer which operates with ultraviolet (UV) and visible (Vis) radiation. This experiment is illustrated in the following as well.

3.1 Gas Injection Methods

In the present work PLP/LIF (cf. section 3.2) and UV/Vis measurements (cf. section 3.3) in the gas phase under slow-flow conditions were carried out. To introduce the gases into the particular setup under controlled conditions different approaches were used.

First studies were carried out with a setup for gas injection, which is called the 'original' setup in this work. However, some aspects of this approach turned out to be problematic (cf. chapter 4). Hence, the gas injection setup was modified. In this section both approaches, the original and the revised one, are introduced. The crucial differences and the advantages of the modification are discussed in chapter 4.

Moreover, different techniques for the gas supply were used. On the one hand, gas mixtures were prepared in high-pressure gas cylinders following the procedures of former works (see e.g. refs. [64–66]). On the other hand, a bubbler cylinder suitable for pressures above atmospheric pressure was developed in this work. Both techniques are also described in this section.

3.1.1 Original Setup

A schematic drawing of the original approach for gas injection is shown in figure 3.1.

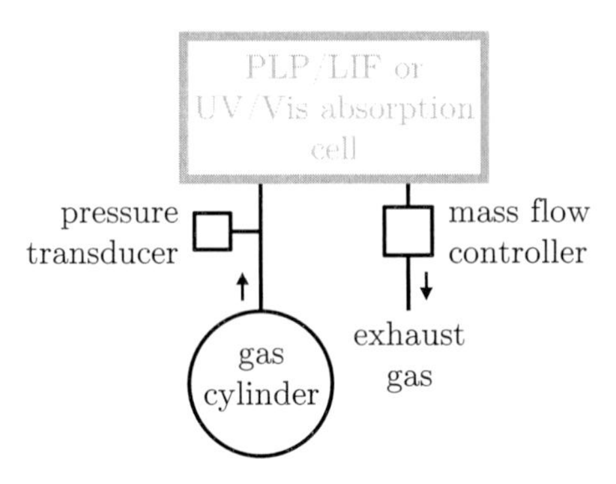

Figure 3.1: Schematic drawing of the original setup for gas injection.

All components, i.e. reactant, radical precursor and bath gas, have to be premixed in one gas cylinder (cf. subsection 3.1.3). The gas cylinder is directly connected with the reaction cell by tubes of stainless steel (6 mm/4 mm, 420 bar or 1/4 in/3/32 in, 4000 bar). Consequently, with this approach the pressure in the cell is determined by the current pressure in the gas cylinder. It is measured with a pressure transducer (tecsis, 3382, <600 bar) before the cell. For a precise regulation of the gas flow, a mass flow controller (Bronkhorst, EL-FLOW, 0.2-10.0 slm) is mounted behind the reaction cell.

Before starting the experiment, the valve of the gas cylinder is opened and the gas flow is set to an appropriate value. After the apparatus is purged a few minutes the measurement can be initiated.

3.1.2 Revised Setup

The revised gas injection setup is schematically depicted in figure 3.2.

In this approach reactant and precursor are provided in different gas supply units. Here, not only mixtures with the bath gas in gas cylinders (cf. subsection 3.1.3) but also the high-pressure bubbler technique (cf. subsection 3.1.4) can be applied. Both methods can be combined freely for the reactant and precursor supply. For an independent regulation of both concentrations, an additional gas cylinder with pure helium is connected in a third line. The gases from the different supply units are injected with three mass flow controllers (Bronkhorst, EL-FLOW, 0.2-10.0 slm) and mixed before the reaction cell with a simple cross connection. For the gas lines, classical rigid tubes of stainless steel (6 mm/4 mm, 420 bar or 1/4 in/3/32 in, 4000 bar) as well as flexible hoses (Fitok, stainless steel, 1/4 in, 213 bar or Swagelok, PTFE, 3/16 in, 206 bar) were used. The pressure in the system is regulated with a backward pressure controller (Bronkhorst, EL-PRESS, 2-10 bar). Additionally, a pressure transducer (tecsis, 3382, <600 bar) is mounted before the particular cell in order to monitor a possible pressure drop in the system.

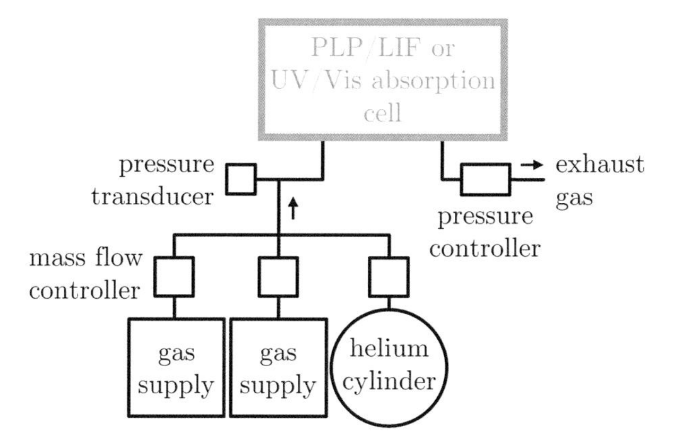

Figure 3.2: Schematic drawing of the revised setup for gas injection.

After opening all gas supply units and regulating the individual gas flows and the pressure in the reaction cell properly, the setup has to be purged a few minutes. Subsequently, the experiment can be started.

3.1.3 Preparation of Gas Mixtures in Cylinders

In principle, the procedure of gas mixture preparation in this work was adopted from former works (see e.g. [64–66]). A schematic drawing of the setup is shown in figure 3.3.

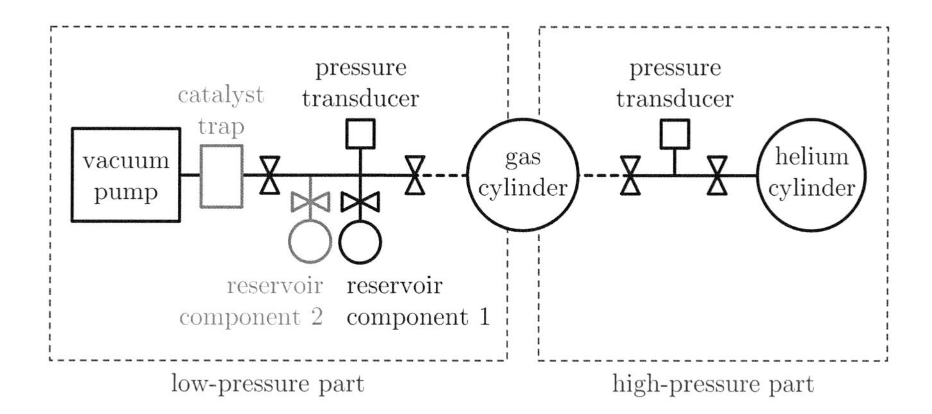

Figure 3.3: Schematic drawing of the setup for mixture preparation in gas cylinders. The gray components were only used in the study of the system DME + OH (cf. chapter 5).

The apparatus can be divided into a low- and a high-pressure part. The parts are separated from each other and possess both a connection for the gas cylinder in which the gases are filled. Two different types of gas cylinders were applied: on the one hand electropolished stainless steel test gas cylinders (Messer Griesheim, 40 l, <300 bar), which are especially suited for corrosive gases, and on the other hand conventional test gas cylinders (Air Liquide, 50 l, <300 bar). The application of both types of cylinders was tested with all different compounds used in this work and generally turned out to be suitable. Nevertheless, due to the high corrosivity of HNO_3, the application of the specialized electropolished stainless steel cylinder is recommended for a mixture preparation involving HNO_3.

First, the cylinder is connected to the low-pressure part. Here, all minor components, i.e. reactant and/or precursor, are filled in. If the components are liquid under standard conditions, an amount is filled in a flask and linked with this part of the setup. Before application, the liquid is carefully degassed. For the use of a gaseous compound the particular gas cylinder is connected.

Afterward, the whole mixing line including the test gas cylinder is evacuated with a rotary vane pump (Alcatel, 2021 SD) to a pressure of approximately $1 \cdot 10^{-3}$ mbar. At the beginning of this work a catalyst trap (Pfeiffer Vacuum, URB 025) was installed between the vacuum pump and the mixing line. Because a leakage appeared, this device was removed after some time. No substitute was inserted, as the operation of the setup is also possible without. However, to avoid a frequent cleaning of the setup due to contamination with the evaporating oil from the pump, which is usually associated with sealing problems, the application of an oil trap is recommended. The leak rate was examined before each mixture preparation and never exceeded $8.4 \cdot 10^{-4}$ mbar l s^{-1}.

When the tightness of the apparatus is ensured the valve to the reservoir is carefully opened and the particular component is expanded into the setup. In the case of liquid substances, the vaporization can be facilitated with the help of a water bath at room temperature. The pressure rise is observed manometrically with a pressure transducer (MKS, Baratron 626B, <20 mbar). When the desired pressure(s) of the minor component(s) are filled in all valves are closed and the test gas cylinder is separated from the low-pressure part of the setup.

Then, it is connected to the high-pressure part. After purging the joining tube with helium, the connection screw is tightened while the valve of the helium cylinder stays open. Now, the test gas cylinder is opened and the current pressure is observed with a pressure transducer (MKS, Baratron 750B, <200 bar). The bath gas is filled in until the target pressure is reached.

A potential shift of the zero point of both pressure transducers is determined during each mixture preparation by means of the lowest pressure in the evacuated low-pressure part and the atmospheric pressure in the relaxed high-pressure part. The measured filling pressures are corrected by this shift.

After preparation, the mixtures are stored for at least 12 hours. This allows homogenization of the compounds inside the cylinder. The question, if this storage time is enough, has already been subject of a discussion in the literature [67–69]. In this work, the dependence of the measured rate coefficients on the age of the gas mixtures was tested and could be excluded for all measurements. Moreover, the homogenization process could be observed indirectly in the UV/Vis absorption measurements of a freshly prepared HNO_3 mixture (cf. section 4.3). It could be estimated from these measurements that, in good approximation, the mixture is homogenized after 12 hours.

When nitric acid is applied as an OH precursor, it is recommended to synthesize the substance freshly before a gas mixture is prepared (cf. chapter 4). The general procedure of the synthesis is taken from the literature [70]. Degassed sulfuric acid is added to dried potassium nitrate under ice-cooling. Nitric acid is isolated as reaction product in a direct vacuum distillation, freezing it with liquid nitrogen. The ice bath is substituted by a water bath at room temperature after a few minutes. The synthesis ran for a maximum of one hour.

3.1.4 Bubbler Technique for High Pressures

The bubbler technique, also referred to as saturator, is a common method for a direct supply of liquid components in the gas phase and frequently used for kinetic measurements with slow-flow reactors (see e.g. refs. [10, 71, 72]). Typically, the applied devices are simple pyrex washing bottles. Thus, they are only suitable for pressures below atmospheric pressure.

In this work a pyrex bubbler for high-pressure measurements was developed [73] and applied in the study of the system HNO_3 + OH (cf. chapter 4). It can be separated into the bubbler section, the cooling unit and the high-pressure housing. Figure 3.4 shows a drawing of the whole apparatus.

The bubbler unit consists of a pyrex lower part and a stainless steel upper part connected with a flange connection. It is sealed with an FKM O-ring and a clamp ring. Two tubes of stainless steel lead away from the upper part. The first centered one passes though, while the other one is fixed at the side and ends at the junction. Both tubes are connected with the housing via screw joints sealed with PTFE sealants. The inner end of the centered tube is connected via a screw joint with a pyrex gas distribution tube, which extends into the lower part of the bubbler. Hence, all elements which are in contact with the liquid are made of pyrex. The application of an upper part made of pyrex turned out to be delicate, as the glass tubes are too fragile to sustain the tensions resulting from the screw connections with the housing and the contact with the cooling.

A cooling system is installed for the regulation of the vapor pressure of the liquid. For this purpose, a copper tube, through which a coolant is streamed, is wrapped helically around a massive brass block. The lower part of the bubbler can be placed into a hollow inside this block. Thermal conductance paste ensures an optimal thermal transfer. The copper tubes

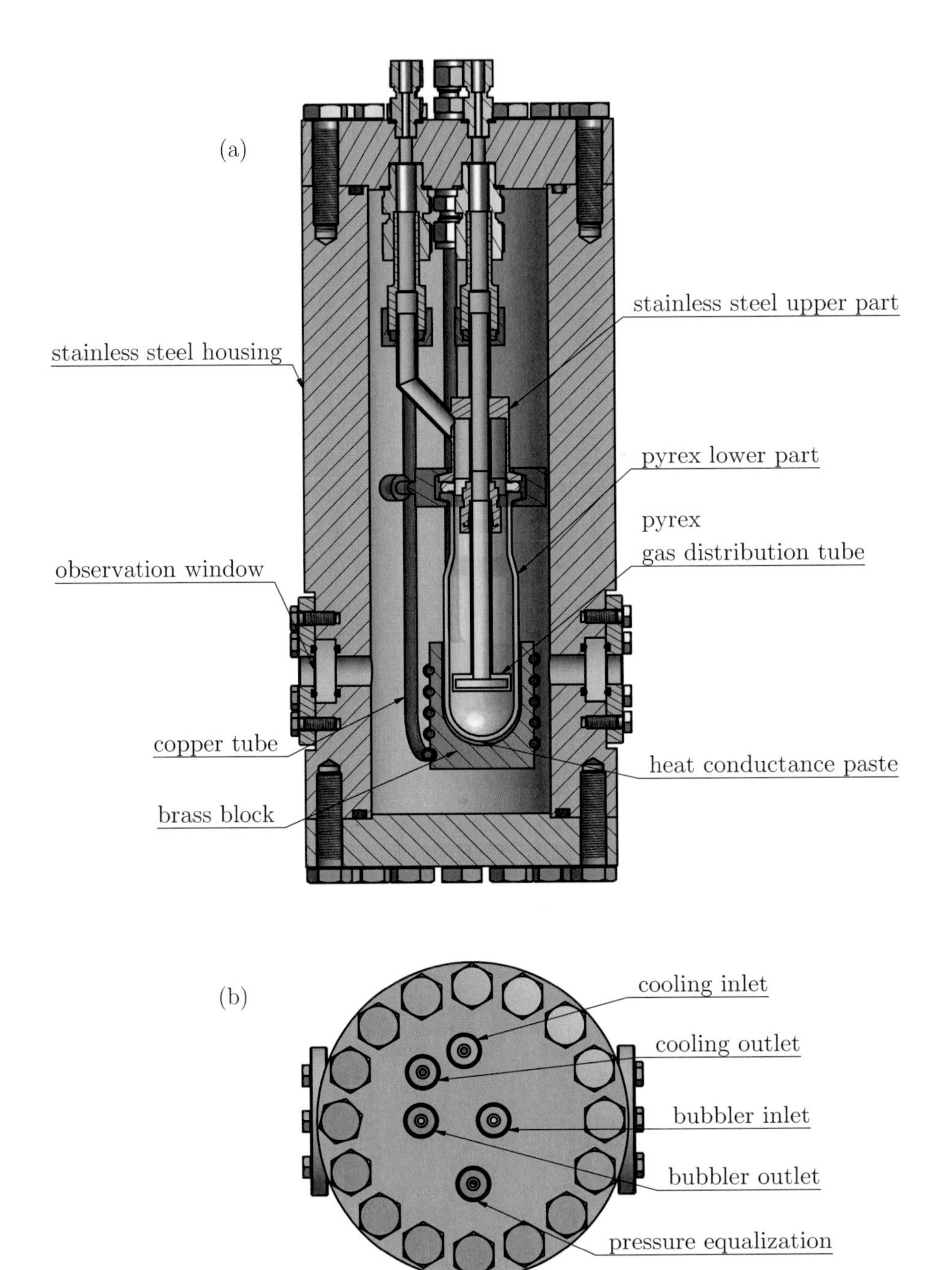

Figure 3.4: Drawing of the high-pressure bubbler apparatus [73]: (a) cross sectional view, (b) top view.

are connected with the housing and with the cooling circuit (Colora, KT40K) outside the housing via screw joints.

The housing of the apparatus is made of stainless steel (material number 1.4571). It has a cylindrical shape with a wall thickness of 3.2 cm. The 30 cm high tube is closed on both sides with flanges of 15 cm diameter. The flanges are sealed with FKM O-rings and each is fixed with 16 screws. Opposite to each other two quartz glass windows are embedded in the tube, which are sealed with FKM O-rings and flanges as well.

To allow a pressure equalization in the volume which surrounds the bubbler part inside the apparatus an additional gas inlet is situated in the housing. It is connected directly with the gas inlet into the bubbler part outside the housing. Moreover, the inlet tubes are linked with the gas outlet tube. All gas accesses can be closed by valves outside the housing. Valves are also mounted between the bath gas supply and the gas inlet tubes and at the connection tube between gas inlet and outlet. In the first case, a needle valve is applied. The pressure inside the bubbler is regulated by the pressure transducer of the bath gas cylinder or with a forward pressure controller (Bronkhorst, EL-PRESS, 20-100 bar), which is mounted between the bath gas cylinder and the needle valve upstream the bubbler apparatus.

After filling in the liquid into the bubbler and closing the bubbler unit, the top flange, on which the bubbler unit is mounted, is put on the hosing cylinder. The screws have to be tightened stepwise and crosswise with a torque spanner with 30 to 40 Nm.

Before putting the bubbler into operation, all gas access valves as well as the connection valve between gas inlet and outlet have to be opened. The needle valve which separates the gas inlet from the bath gas supply has to be closed. Then, the bath gas cylinder is opened and the pressure is regulated to the desired value. Afterward, the needle valve is opened carefully. A fast opening of the needle valve leads to a fast pressure rise inside the bubbler. In this case, the pressure equalization can occur too slow and the pyrex flask can burst or the bubbler tubes can be pushed out of the screw joints. After building up the pressure inside the apparatus and regulating the gas flow, the valve between gas inlet and outlet is closed carefully. Thus, the bath gas is redirected through the bubbler and the bubbling inside the flask is started. It can be observed through the windows in the bubbler housing if the cooling block is not used.

3.2 Pulsed Laser Photolysis/Laser-Induced Fluorescence

In principle, a PLP/LIF experiment starts with the photolysis of a radical precursor by laser light. The generated radicals then react with molecules which are present in the gas phase. With a second laser these radicals are excited to fluoresce and the emitted radiation, which is proportional to the radical concentration, is detected. The precise control of the time delay between photolysis laser pulse and probe laser pulse permits the determination of the radical fluorescence intensity at a certain point in time after radical generation. By pulsing the whole experiment and changing the time delay between photolysis and probing stepwise, the decay of the radical concentration can be observed and the rate of the reaction can be measured.

The PLP/LIF setup applied in this work can be separated into four parts. First, there is the setup itself composed by the optical and electronic devices. The centerpiece of the setup is the reaction cell, where the reaction of interest takes place. In the present work two different cells were used. They are treated separately as the second part. The third part of the setup is a unit for absorption measurements. It was built up in this work as an alternative determination method of the reactant concentration. The gas injection system is the fourth component of the experiment. It serves the purpose of providing the gas mixture in the reaction cell under the desired conditions. Because it was built up in an analogical way for the UV/Vis measurements (cf. section 3.3), it is described in a separate section (see section 3.1).

Crucial for the applicability of PLP/LIF is the availability of an adequate radical precursor and the possibility of fluorescence excitation of the reacting radical by the provided instruments. The hydroxyl radical (OH), which is the educt radical in all systems studied in this work, is one of the most common radicals investigated with the PLP/LIF technique. Several OH radical precursors are known (cf. section 4.1). Moreover, it possesses a strong fluorescence band in the near UV excitable by common lasers. As a result, even at high pressures an intense fluorescence signal is detectable. The detailed circumstances and conditions of the radical generation and detection are described in this section, too.

Finally, in a detailed error discussion the uncertainties of the rate coefficients determined in the PLP/LIF experiments are evaluated.

3.2.1 Experimental Setup

The PLP/LIF setup is schematically depicted in figure 3.5. The colored components are introduced in separate (sub)sections.

For the photolysis of the precursor an excimer laser (Lambda Physik, Compex 102, KrF, 248 nm) is used. The detection laser is a dye laser (Lambda Physik, Scanmate 2E) operated with Coumarin 153 at approximately 564 nm. It is pumped by another excimer laser

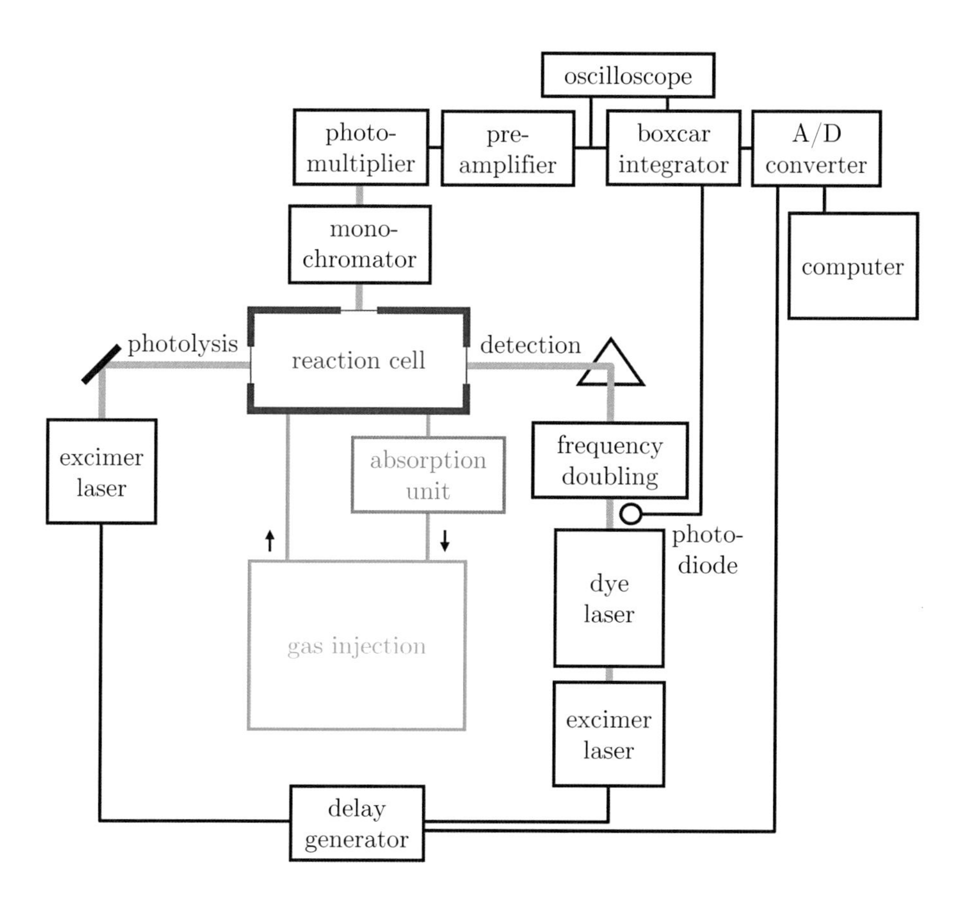

Figure 3.5: Schematic drawing of the PLP/LIF experiment. For a clearer illustration only one of the particular adjustment optics is shown.

(Lambda Physik, Compex 102, XeCl, 308 nm). The excitation wavelength of approximately 282 nm is generated by a BBO-crystal (Radiant Dyes, 220-320 nm) in the corresponding frequency doubling unit. The problem of a slight decalibration of the dye laser wavelength and the detailed dye laser settings are discussed in subsection 3.2.4. The pulse duration of both lasers amounts to approx. 25 ns. Hence, it can be approximated that photolysis and excitation are instantaneous on the time scale of typical chemical reactions in the gas phase.

The laser beams are adjusted on the opposite reaction cell windows in a coaxial propagation with three dielectric mirrors in the case of the photolysis beam, and three quartz prisms in the case of the detection beam. The whole free window surface is illuminated by the photolysis laser, while the excitation laser beam exhibits a diameter of approx. 1 mm. It was always adjusted onto the center of both windows.

The fluorescence is recorded through the window perpendicular to the laser axis. Here, wavelengths other than the detection wavelength are first filtered out by a monochromator (Carl Zeiss, M4 QIII), which provides a tunable slit for the adjustment of the bandwidth. For all experiments carried out in the present work, non-resonant detection was chosen, as the scattering light not only of the photolysis laser but also of the detection laser can be minimized and the signal-to-noise ratio can be improved in comparison to a resonant detection. The particular monochromator settings which were chosen in the different studies are given in the corresponding chapters. Behind the monochromator a photomultiplier tube (PMT) (Hamamatsu, R212) is mounted detecting the passing radiation.

The PMT signal is amplified with a fast preamplifier (Stanford Research Systems, SR250) and integrated with a boxcar integrator (Stanford Research Systems, SR205). The integration gate is regulated manually on the fluorescence signal by observing both signals on an oscilloscope (LeCroy, 9361). Generally, the whole fluorescence peak was included in the integration of the PMT signal. Then, the obtained analog signal is digitized by an analog-to-digital (A/D) converter (National Instruments, USB 6229) and recorded by the computer with the help of a measuring program written by Hetzler in LabView [74].

For the time control of the experiment a delay generator (Stanford Research Systems, DG535) is applied. Its internal trigger defines the starting point t_0 of each experimental run. The trigger of both excimer lasers and the reset of the A/D converter are provoked with a given time delay relative to t_0, which can be controlled with the measuring program.

The integration gate of the boxcar integrator is not directly triggered by the delay generator but by a photodiode, which sends a trigger signal each time it detects a laser pulse of the dye laser. This procedure avoids a jitter in time of the integration gate relative to the fluorescence signal which is observable when the integration gate is directly controlled by the delay generator. It arises from the slightly varying time delay between trigger pulse and electric discharge in the pumping excimer laser. Due to this, the point of time of the fluorescence signal also varies a little bit relative to the trigger of the delay generator.

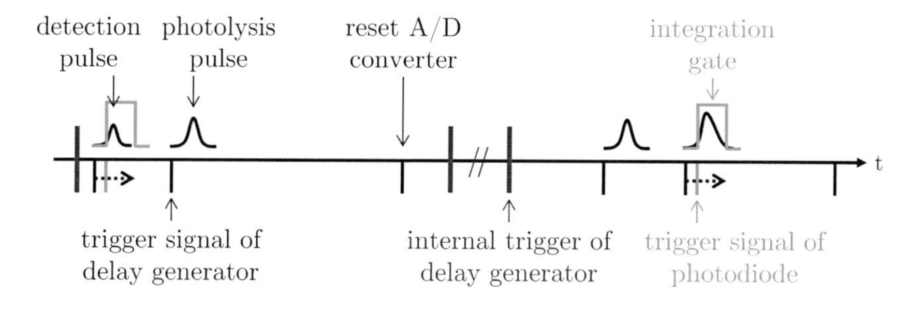

Figure 3.6: Schematic diagram of the chronology of the PLP/LIF experiment.

The chronology of the experiment is illustrated in Figure 3.6. At the beginning of a measurement the time delay between photolysis and detection trigger is set to a negative value. Hence, no OH radicals are present in the reaction cell when the detection laser beam passes through and the recorded values of the integrated PMT signal represent the zero baseline. The photolysis/detection time delay is then automatically increased with every trigger pulse in steps defined in the measuring program. Once the detection pulse falls back in time behind the photolysis pulse, OH radicals are present for excitation by the detection laser if the step size is chosen properly. By further changing the trigger of the detection pulse relative to the photolysis pulse, the detected OH concentration again decreases as the desired chemical reaction takes place. When the time delay between photolysis and probe pulse is so long that all OH radicals are consumed by the reaction a complete concentration-time profile is recorded containing the kinetic information of interest.

The signal-to-noise ratio can be increased by recording several measuring points at a given time delay and taking the mean value. Such an averaging is also implemented in the program, while the number of included single values can be defined by the user.

The particular wavelengths, laser fluences, trigger frequencies and averaging values which were chosen for the investigation of the individual systems are specified in the corresponding chapters.

3.2.2 Reaction Cells

The reaction cells applied in the present work are both suitable for measurements at pressures above 1 bar. The first cell is equipped with a cooling system, which enables experiments below room temperature. The second heatable one serves for measurements at temperatures higher than room temperature. Both cells were built previously and have already been described in detail in former works [64, 75].

The coolable reaction cell was first described in the work of Fulle [75]. Figure 3.7 shows the schematic drawing of the cell and the surrounding vacuum chamber.

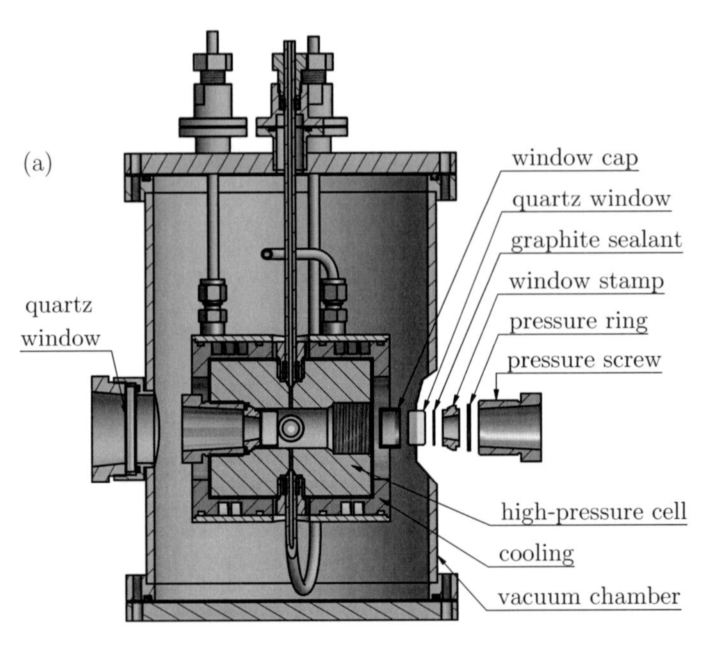

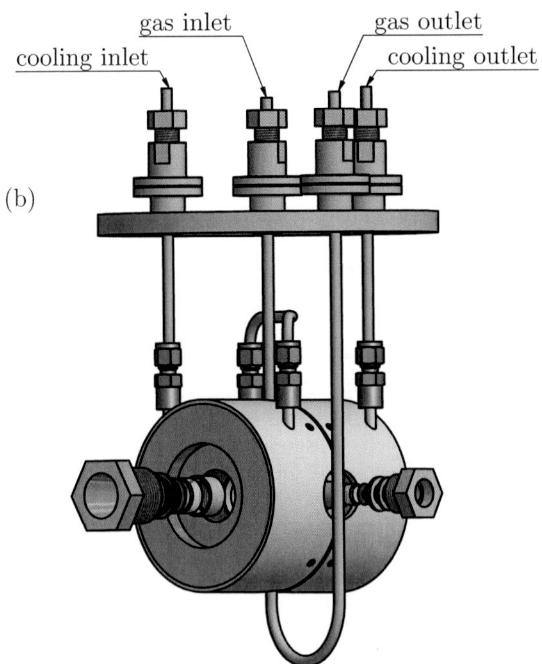

Figure 3.7: Drawing of the coolable high-pressure cell: (a) cross sectional view with vacuum chamber, (b) three-dimensional view of the cell and the top of the vacuum chamber.

The cell itself is a stainless steel cylinder with four drill holes for the optical access. Two bigger windows (Heraeus, Suprasil, 20/10) are situated on the bases and two smaller ones (Heraeus, Suprasil, 12/8) on the side faces, each opposite to each other. The windows, which are made of quartz glass, are fixed on the polished bottom side of the corresponding stamps by caps. To avoid breaking, they have to be chamfered. A graphite ring (thickness 0.5 mm) between window and stamp serves as sealant. The stamps are fixed in the cell body with pressure screws and pressure rings. The sealing between cell body and stamp is achieved by the conical shaped planes.

The laser beams are adjusted on the two smaller windows, while fluorescence is detected through one of the bigger windows. The fourth window is not needed for a conventional PLP/LIF experiment and is therefore blocked.

The cylindrical cavity in the center of the cell contains the reaction gas. The length and the diameter of the cavity amount to 16.0 mm and 23.4 mm, respectively, which gives an inner gas volume of 6.88 cm^3. The diameter of the free part of the smaller windows amounts to 4 mm. As the reaction only takes place in the volume irradiated by the photolysis laser, a reaction volume of 0.20 cm^3 results. Due to the minor diameter of the probe laser beam, the gas volume in which the radicals are observed is even smaller. For the gas inlet and outlet high-pressure gas tubes are directly mounted in the cell. They pass through the top of the vacuum chamber and can be connected outside with the gas injection unit.

The reaction cell is surrounded by a cooling jacket. It consists of two parts, each made of a bronze body and a stainless steel closure. In the bronze body a helical channel is embedded, through which the coolant streams. The sealing between the body and the closure is achieved with O-rings. The cooling parts are put on the side faces of the cell cylinder, while the window accesses are left open. The coolant streams into and out of the jacket through conventional high-pressure gas tubes, which also pass through the top of the vacuum chamber. Outside, the cooling system is mounted. Two different systems are available for this purpose. On the one hand, a commercial cryostat (Colora, KT90S) operated with ethanol can be applied. With this apparatus the temperature can be adjusted and held constant without difficulties. However, it can take some time until the desired temperature is reached and the minimal achievable cell temperature is limited to approx. 250 K. On the other hand, a cooling system with nitrogen built up in a former work [75] is available. Here, liquid nitrogen is filled in a reservoir vessel, which is closed so that the pressure in the vessel rises. Then the valve leading to the cooling jacket of the cell is opened and the cell is cooled by cold nitrogen gas or with liquid nitrogen. With this cooling system, the temperature decrease is very fast and very low temperatures can be achieved. A disadvantage though is the fact that no constant temperature can be adjusted. Consequently, it has to be regulated very often.

Reaction cell and cooling are incorporated into a vacuum chamber, which has an optical access at every position of the cell windows. The top and the bottom of the cylindrical

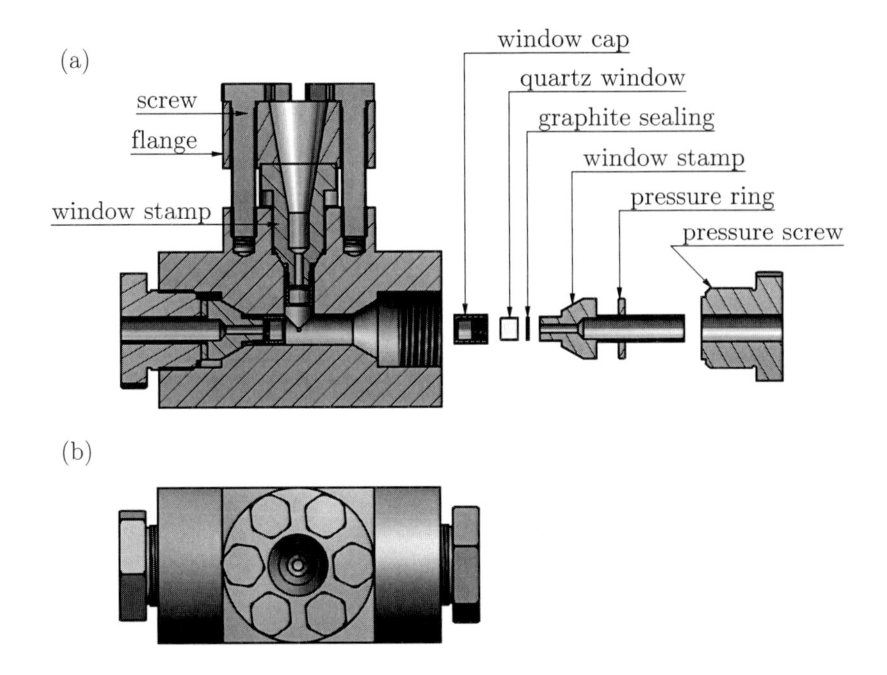

Figure 3.8: Drawing of the heatable high-pressure cell: (a) cross sectional view, (b) top view.

chamber are removable and are closed with flanges and sealed with O-rings. The inner volume of the chamber is evacuated with a rotary vane pump (Pfeiffer, Duo 10M). This procedure avoids condensation of water on the windows of the cooled reaction cell.

An additional pre-cooling made of stainless steel can be put on the inlet capillary and can also be connected with the cooling system. It was constructed by Zügner [76] to reduce the high temperature difference between gas inlet and outlet inside the reaction cell at low temperatures. However, it has not been tested yet.

The reaction cell for high temperature measurements was constructed by Striebel [64]. A schematic drawing of the cell is depicted in figure 3.8.

The body of the cell is made of highly heat resisting stainless steel. Three chamfered quartz windows (Heraeus, Suprasil, 10/10) are integrated in a T-shape with the help of window stamps in the same manner as in the coolable high-pressure cell. The stamps of the opposite windows are fixed in the body with pressure screws, whereas the one perpendicular to this axis has a flange closure.

The inner gas space has a cylindrical shape. The gas inlet and outlet are situated above and below the plane of projection in figure 3.8 (a). The distance between the two opposite windows, as well as the diameter of the cavity, amounts to 14 mm. Consequently, the gas volume of the cell is 2.16 cm^3. With the diameter of the free part of all windows, which amounts to 5 mm, a reaction volume of 0.27 cm^3 is obtained. Again, the detection volume is even smaller because of the smaller diameter of the probe laser beam.

For the heating of the cell, resistance heating wires (Les cables de Lens, Pyrolyn-M, 16 Ω/m) are wrapped on a brass jacket, which surrounds the cell only sparing the windows. In the course of this work a pre-heating was installed by Zügner [76]. It consists of a heating cord (Isopad, TN 7050), which is wrapped around the inlet capillary to a length of approx. 55 cm. With this pre-heating temperature inhomogeneities inside the cell could be eliminated. However, it was only used for the study of the systems DEE(-d10) + OH and DMM + OH. Both heatings are isolated with fiber glass tape. To survey the temperature difference inside the reactor, the temperature is measured by two thermocouples (eNET Fühlersysteme, Ni/Cr-Ni), which are embedded in the gas tubes of the inlet and outlet. The mean of both temperatures is assumed as reaction temperature. The reference of the thermocouples was changed during this work. For the measurements on the system DME-d6 + OH an ice bath reference was applied. In the following, a cold junction compensation was installed, which was used in all other measurements. This increases the error of the temperature measurement to 1 K, but avoids the effort which is connected with the application of an ice bath.

3.2.3 Absorption Unit

During the present work an absorption unit was integrated in the high-pressure PLP/LIF setup. It serves as an alternative determination method of the reactant concentration. In this work it was only used for the measurements on the system HNO_3 + OH, which are described in subsection 4.5.1.

Centerpiece of the setup is the high-pressure absorption cell. It was constructed by Acalovschi and has been primarily described in ref. [77]. A schematic drawing of the cell is shown in figure 3.9

The absorption cell consists of a cylindrical body made of stainless steel with a height of 20.5 cm and a diameter of 5.9 cm. The quartz glass windows (Heraeus, Suprasil, 16/11) are embedded in each base of the cylinder. They are sealed with graphite rings and mounted in the cell with stamps and pressure screws in an analogous way as in the low-temperature LIF cell (cf. subsection 3.2.2). The diameter of the free window surface amounts to 8 mm. The exact optical path length was remeasured with a high accuracy. A value of 10.514 cm was obtained.

As a UV light source a low-pressure, cold-cathode zinc lamp (UVP, Pen-Ray) is used. It is operated with alternating current (50 Hz), which is why the emission intensity changes periodically with a frequency of 100 Hz. A power supply for a direct current supply was not available.

Other wavelengths than the detection wavelength at 214 nm are filtered out with a

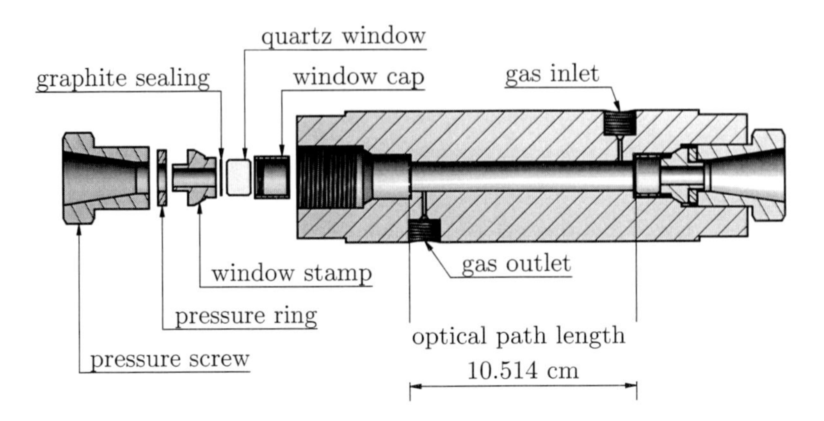

Figure 3.9: Cross sectional drawing of the high-pressure absorption cell applied in the absorption unit in the PLP/LIF setup.

monochromator (Carl Zeiss, M4 QIII). The width of the slit is set to 2 mm, which corresponds to a band width of 8 nm (fwhm). On the other side of the cell the passing radiation is detected with a PMT (Hamamatsu, R212).

The PMT signal is amplified with a fast preamplifier (Stanford Research Systems, SR250), digitized with an A/D converter (National Instruments, USB 6229) and recorded by the measuring program [74]. Due to the above mentioned alternating lamp emission, only the maximum values of the recorded signals in periods of 10 ms are stored and the others are deleted.

The unattenuated base intensity I_0 is measured before each PLP/LIF measurement with a constant helium flow through the absorption cell and saved internally in the program. When the PLP/LIF experiment is started the attenuated intensity I is recorded until the measurement is stopped again. All obtained values for I are averaged. From this mean value, the absorbance and the reactant concentration are calculated automatically by the program according to equation 2.10 and written into the output file.

3.2.4 Radical Production and Detection

The photodissociation of HNO_3 excited by UV radiation has been studied extensively in the last decades. A short review is published by Huber [78].

The UV/Vis absorption spectrum of HNO_3 shows two broad absorption bands in the UV range. One intense band ranges from 150 to 240 nm and has its maximum at approximately 183 nm. It is overlapped by another weak band at around 270 nm. [78]

An HNO_3 molecule excited into the latter dissociates exclusively to NO_2 and OH radicals:

$$HNO_3 \xrightarrow{h\nu} OH + NO_2. \tag{R10}$$

Excitation into the absorption band at lower wavelengths leads to a more complex photochemistry. The photolysis wavelength of 248 nm lies on the overlap of the two bands. Here, the quantum yield ϕ for reaction R10 is already slightly below one [79, 80].

However, considering the accessible wavelengths of excimer lasers, a photolysis at 248 nm turned out to be a good compromise for kinetic studies. Here, the generation of significant amounts of reactive by-products can still be neglected, while the cross section is already large enough for an acceptable OH yield [11]. Moreover, the suitability of HNO_3 photolysis at 248 nm for the generation of OH radicals in kinetic measurements has already been proven in numerous studies (see e.g. refs. [10, 81, 82]).

Hence, the quantum yield of reaction R10 in this work is only relevant for the estimation of an upper limit of [OH] validating the pseudo-first order assumption. For this purpose, a simplified approximation of ϕ with a value of one is reasonable.

The number of photons which are absorbed by nitric acid can be calculated with the Beer-Lambert law given in equation 2.10. Assuming that only single-photon absorption takes place, the OH concentration [OH] is given by:

$$[\mathrm{OH}] = \phi \frac{F\left(1 - \exp\left(-\sigma_\lambda(T)[\mathrm{HNO_3}]d\right)\right)}{\mathrm{h}\frac{\mathrm{c}}{\lambda}d}. \tag{3.1}$$

ϕ is the quantum yield, F the fluence of the excitation laser, d the optical path length and λ the excitation wavelength. h represents the Planck constant, while c is the velocity of light. The temperature-dependent absorption cross section $\sigma_\lambda(T)$ of HNO_3 at 248 nm is taken from ref. [83], following the recommendation of ref. [27]. It is given by:

$$\sigma_{248}(T) = 2.00 \cdot 10^{-20} \exp\left(\frac{1.44 \cdot 10^{-3}(T - 298\ \mathrm{K})}{\mathrm{K}}\right)\ \mathrm{cm}^2. \tag{3.2}$$

After dissociation at 248 nm the main excess energy of the photon absorption remains in the NO_2 molecule or is turned into translational energy. Only around 5 % stays in rotational and 1 % in vibrational modes of the OH radicals. [84] Furthermore, the equilibration of the small particles should be fast on the time scale of the chemical reaction at the high pressures applied. Hence, a thermal population of the reacting OH radicals can be assumed.

Laser-induced fluorescence of OH radicals is a well-established method. Thus, for the OH detection in this work the common procedure described in the literature was applied (see e.g. [10, 65, 85, 86]). OH radicals were excited at 281.9 nm, which corresponds to the $A\ ^2\Sigma^+\ (\nu' = 1) \leftarrow X\ ^2\Pi\ (\nu'' = 0)$ transition. The fluorescence was detected non-resonantly

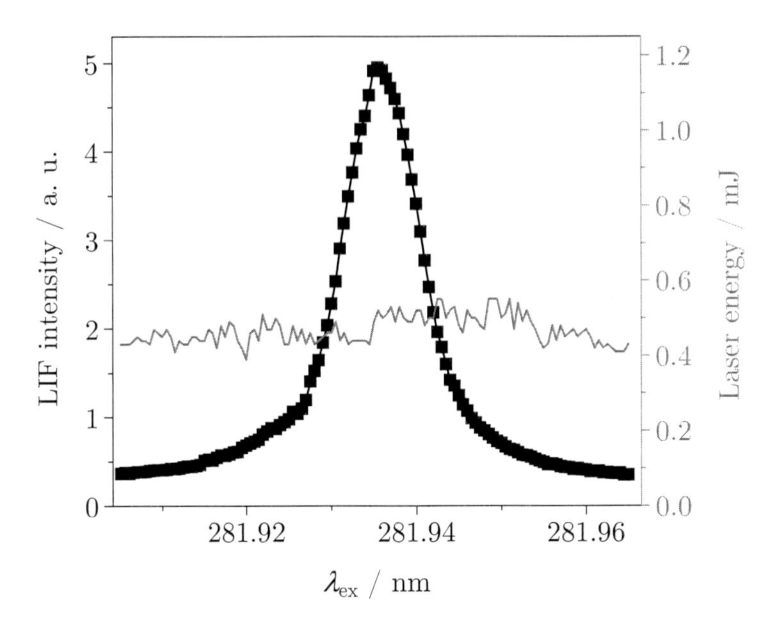

Figure 3.10: Excitation spectrum of OH recorded at 294 K and 10 bar. The plotted excitation wavelength λ_{ex} was directly taken from the dye laser settings without any recalibration.

at (308 ± 4), (308 ± 8) and (316 ± 9) nm [86], respectively.

Because the absorption bands of this transition are very sharp [28, 87], the fluorescence yield is very sensitive on small variations in the excitation wavelength. However, a continuous recalibration of the dye laser output wavelength λ_{out}, set in the dye laser program, is not practicable. Consequently, its adjustment was usually carried out by optimizing the observed fluorescence signal. At the beginning of the experiments of this work the excitation wavelength was adjusted in this way to a value of 281.923 nm. In the course of the present work a slight decalibration of the dye laser output wavelength was observed by means of a decrease in the detected OH fluorescence intensity. Hence, an excitation spectrum of the OH band assigned to the above-named transition was recorded. It is shown in figure 3.10. In the following, the laser output wavelength for OH excitation was set to the obtained maximum at 281.936 nm.

3.2.5 Error Analysis

For the PLP/LIF experiment several potential sources of error have to be considered. Due to the different experimental methods applied, some of them have to be discussed separately, while others are valid in all cases. Hence, an overall error analysis is given here, in which in some points it is distinguished between the different approaches.

The error of the temperature is influenced by several factors. First, the temperature measurement itself exhibits a statistical error. However, in the case of the ice bath reference, which was applied in the study of DME-d6 + OH, it is very small and can be neglected. In the case of the cold junction compensation the uncertainty is specified by the producer to 1 K. For the PLP/LIF measurements in this work this is a tolerable error. In addition, the temperature difference between gas inlet and outlet contributes to the statistical error of the temperature. During this work it was minimized by the installation of a pre-heating in the heatable PLP/LIF cell. Nevertheless, the detection zone, which is located in the center of the cell cavity, is very small. Thus, the temperature gradient in this area is assumed to be insignificant and the actual reaction temperature is well described with the mean of the inlet and outlet temperature. A third influencing factor is the temperature change during one series of measurements, as one bimolecular rate coefficient is obtained from these measurements. This difference varies considerably in the different series. In most cases the doubled standard deviation from the averaging was at or above 1 K. Hence, this value was taken for the quantification of the error. In a few series of measurements the doubled standard deviation of the mean temperature lay below 1 K. In this case the error was rounded up to 1 K to account also for the uncertainty of the cold junction compensation.

Regarding the error of the measured rate coefficients, numerous aspects have to be considered. First, the validity of the pseudo-first order assumption is of particular importance. It was examined by the determination of an upper limit of the ratio [OH]/[reactant]. Moreover, the pseudo-first order rate coefficients were plotted against the reactant concentration and the linearity was checked. A detailed discussion of these points is given in the corresponding chapters of the particular systems. In all cases the assumption of pseudo-first order conditions could be verified.

Another possible error source is the removal of OH radicals by competitive processes. Here, in most points the different approaches for gas injection have to be discussed separately.

In the case of the original setup for gas injection, which was applied in the measurements on the system DME-d6 + OH, the reactant concentration cannot be varied independently. As a result, the bimolecular rate coefficients are obtained by dividing the pseudo-first order rate coefficients by the reactant concentration. If another OH removing process contributes significantly to the observed OH decay, it therefore causes a systematic error. Consequently, such possible influences have to be evaluated carefully.

The competitive OH removing processes can be divided into physical and chemical processes. One physical process is the diffusion of the OH radicals out of the detection volume. Its influence can be evaluated by the root mean square distance $\sqrt{\overline{x^2}}$ covered by a diffusing particle. It is given by the relation

$$\sqrt{\overline{x^2}} = \sqrt{2Dt}\,, \tag{3.3}$$

where D represents the diffusion coefficient and t the diffusion time [88]. With the diffusion coefficient of H radicals in helium at 500 K [89] and the maximum reaction time in the measurements on the system DME-d6 + OH of $2 \cdot 10^{-5}$ s the upper limit for $\sqrt{x^2}$ was roughly estimated. A value of 0.14 mm was obtained, which is less than 3 % of the diameter of the free window surface. In addition, the volume in which the OH radicals are detected is significantly smaller than the whole reaction volume. Thus, the number of OH radicals diffusing into and out of the detection volume should be the same in good approximation.

Moreover, a transport of the OH radicals results from the adjusted gas flow. It is highly dependent on the flow conditions inside the reaction cell. A turbulent behavior is assumed for the specific characteristics of the setup. The influence was examined empirically by the investigation of a potential dependence of k_{2d} on the gas flow. No correlation was observed.

Apart from these physical processes, OH side reactions also have to be considered. One of them is the reaction of the radical precursor with OH. The proportion of HNO_3 + OH on the measured pseudo-first order rate coefficient in the study of reaction DME-d6 + OH was calculated by means of the high-pressure limit of k_1 from ref. [10]. It is always less than 0.6 %. Hence, the error caused by HNO_3 + OH in the study of the system DME-d6 + OH is negligible.

Another possible reason for OH removal is the reaction of impurities of the OH precursor. It could be shown that significant amounts of HNO_3 decompose to NO_2 inside the cylinders during the storage of the mixtures (cf. chapter 4). Because the reaction of NO_2 with OH radicals is very fast [27], this side reaction can influence the accuracy of the measurements significantly. As the exact concentration of NO_2 in the gas mixtures is not known, the quantification of this error in the particular measurements is difficult. Nevertheless, with different efforts it could be shown that the error of k_{2d} caused by NO_2 impurities under the applied conditions is negligible within the error range of the experiment. Firstly, the rate coefficient of the undeuterated system was measured in a previous work [14]. The remarkable agreement with the numerous rate coefficients reported in the literature proves the reliability of the method in this case. Secondly, the difference between the measured and fitted rate coefficient was plotted against the age of the gas mixture. No correlation was noticeable. Thirdly, the rate of the side reactions of the precursor and its impurities with OH was examined with gas mixtures containing only HNO_3 and helium. Although the exact influence of NO_2 + OH on the measured rate coefficient k_{2d} cannot be quantified this way, a rough estimation of the OH lifetime could be achieved and the DME-d6 concentration could be chosen adequately.

With the revised setup for gas injection, all these OH removing processes do not cause a systematic error in the bimolecular rate coefficients anymore. Here, pseudo-first order rate coefficients at different reactant concentrations are determined in one series of measurements with all other parameters, like pressure and precursor concentration, held constant. Thus, the amount of removed OH radicals due to transport processes and side reactions of the precursor molecule and its impurities is the same for all measurements in one series. The bimolecular rate coefficient is obtained from the slope of the plot of the pseudo-first order rate coefficients against the reactant concentration. All above-named OH removing processes contribute to the intercept and do not influence the slope.

However, the offset in the pseudo-first order plots increases significantly with the age of the HNO_3 gas mixtures due to the increasing amount of NO_2. Moreover with the decomposition of HNO_3, the concentration of the effective fluorescence quencher water in the gas mixture increases. Both aspects lead to a decreased signal-to-noise ratio in the LIF intensity-time profiles measured with HNO_3 mixtures which are stored for a longer time period. Consequently, attention was paid that only HNO_3 mixtures with a low age were applied.

The influence of the recombination of two OH radicals is of relevance for both approaches of the experimental setup. It could be excluded in all studies of this work by varying the fluence of the photolysis laser. With this variation, the initial OH concentration was altered (cf. subsection 3.2.4). As the rate of the recombination reaction depends quadratically on the OH concentration, the measured rate coefficients should correlate with the laser fluence if the OH recombination has an influence. Such a correlation was not observed. The range of the variation is given in the corresponding chapters.

Another important factor which contributes to the error of the bimolecular rate coefficient, is the error of the reactant concentration. In the case of the studies of the systems DME-d6 + OH, DEE(-d10) + OH and DMM + OH the reactant concentration was determined from the filling pressures in the mixture and the current pressure in the reaction cell. Here, the errors of the pressure measurement during mixture preparation lead to systematic errors of the reactant concentrations in the particular mixtures. Consequently, the difference between the measured rate coefficients and the corresponding values from the best fit of the Arrhenius expression was plotted against the number of the mixture. No significant dependence was observed.

Moreover, the error of the pressure measurement in the reaction cell has an influence. Furthermore, the ratio $f_\mathrm{R}/f_\mathrm{overall}$ accounts as an additional factor in the calculation of the reactant concentration when the revised setup for gas injection is used. Hence, the uncertainties of the flow controllers also influence the overall concentration error in this case. However, all these influencing factors are assumed to contribute statistically and are thus reflected in the scatter of the pseudo-first order rate coefficients.

In the case of the study of the system HNO_3 + OH the reactant concentration was determined in absorption measurements with the help of equation 2.10. Here, the error of the absorption cross section from the literature [83] and of the optical path length are assumed to be negligible. The temporal stability of the baseline of the PMT signal was examined and turned out to be poor on a time scale of hours. Thus, the base intensity I_0 was determined before each measurement. In addition, all single values for the reactant concentration which were recorded during one measurement were read out separately and plotted against the time. Because no systematic drift was observed in these plots, a systematic error caused by a baseline drift can be excluded. Hence, the most important uncertainty of these concentration measurements results from the scatter of the PMT signal. Due to the high averaging number of single measurements, it is assumed to be small.

The statistical error of the rate coefficients is also influenced by several parameters regarding fluorescence excitation and detection. The scatter of the photolysis laser energy leads to a scatter of the concentration of the produced OH radicals, which increases the scatter in the measured LIF intensity-time profile. Thus, an effort was made to reach a high stability of the photolysis laser, e.g. by changing the thyratron. However, this influence on the statistical error is still assumed to be significant. Due to the application of the saturated LIF technique (cf. subsection 2.1.4), the scatter of the excitation laser energy influences the signal-to-noise ratio of the measurement marginally as long as the scattered laser radiation can mostly be filtered out.

In contrast, the uncertainty of the time measurement is another source of statistical error in the measured profiles. With 50 ps, the uncertainty of the delay generator is negligible. However, a small scatter on the time scale of nanoseconds arises from the differing time delays between trigger pulse and electrical discharge in the excimer lasers varying from pulse to pulse. In addition, the scatter of the PMT signal contributes to the overall statistical error. It can be minimized by the choice of an adequate potential of the power supply. Nevertheless, the scatter of the PMT signal contributes significantly to the signal-to-noise ratio of the measured LIF intensity-time profiles. The influence of the uncertainty of the different electronic devices for the recording of the LIF intensity is assumed to be negligible.

With this discussion the multitude of influencing factors on the overall error of the obtained rate coefficients becomes obvious. Due to the possibility of small systematic errors in the measurements on the system DME-d6 + OH with the original setup for gas injection, it cannot be assumed that all uncertainties of the experiment are reflected by the scatter of the single values from one series of measurements. Hence, the approach for error analysis of former works was followed and for the error of the bimolecular rate coefficient k_{2d} the estimate of 30 % was taken [64, 65].

The revision of the gas injection setup served amongst other reasons for the elimination of the most prominent systematic error sources. Hence, it is assumed that the scatter of the measured pseudo-first order rate coefficients characterizes the overall error of the experiment. As a result, for all measurements with the revised setup for gas injection, the error of the bimolecular rate coefficients was described with the doubled standard deviation of the slope obtained from the best linear fit.

3.3 UV/Vis Absorption

For the identification and quantification of reactive impurities of the OH precursor nitric acid (cf. chapter 4), a UV/Vis absorption setup was built up. Here, a commercial UV/Vis/NIR spectrometer (Varian, Cary 500) was applied. Static measurements at pressures below atmospheric pressure as well as measurements under slow-flow conditions at high pressures were carried out. Both experiments are described in this section.

3.3.1 Static Experiments

The setup of the static experiment is illustrated in figure 3.11.

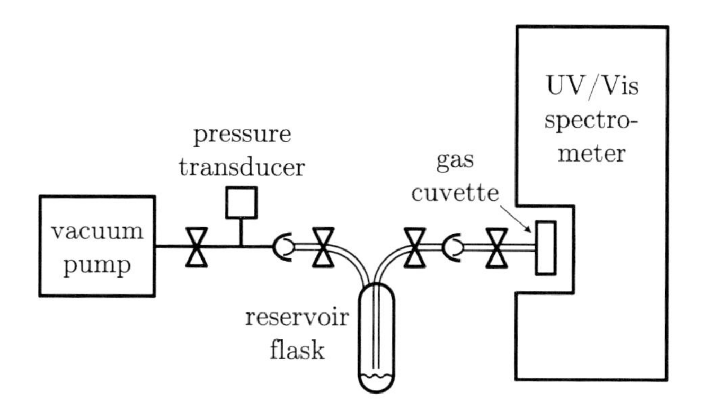

Figure 3.11: Drawing of the static UV/Vis setup.

Here, a glass gas absorption cell with a cylindrical shape and one inlet is applied. The quartz glass windows are glued to the glass body on each base of the cylinder. The optical path length amounts to 11.5 cm. The access tube is connected outside the spectrometer by a spherical joint with the inlet of the reservoir flask. Both, the absorption cell and the reservoir flask, can be closed with valves. The outlet of the reservoir flask is connected to a T-piece, at which a vacuum pump (Pfeiffer, Duo 10M) and a pressure transducer (tecsis, 3382, <600 bar) are mounted. The vacuum pump can be disconnected from the setup by a valve.

3.3.2 Slow-Flow Experiments

For the measurements under slow-flow conditions at pressures above 1 bar, a high-pressure absorption cell was constructed in this work. Figure 3.12 shows the drawing of the cell.

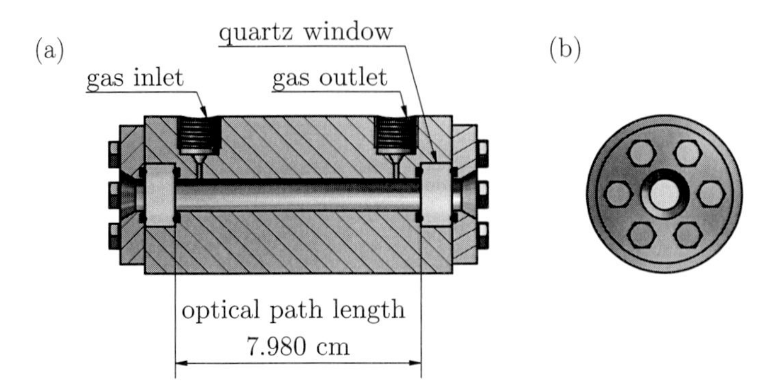

Figure 3.12: Drawing of the slow-flow UV/Vis absorption cell: (a) cross sectional view, (b) side view.

The optical accesses are embedded in each base of the cylindrical cell. Quartz glass windows (Heraeus, Suprasil, 20/10) are inserted and sealed with flanges and O-rings. The optical path length amounts to 7.980 cm. The gas inlet and outlet are located abreast at the cylinder side. Here, the gas tubes of the gas injection unit are mounted with conventional screw joints. The absorption cell is fixed inside the spectrometer with a special mount and the sample space is shaded with a special covering. Both parts were constructed particularly for this application. The original and the revised approach were used for gas injection. The setups are described separately in section 3.1.

4 Studies on the Purity of HNO_3 and Revision of the PLP/LIF Setup

4.1 Introduction

Due to the high relevance of OH radicals for gas phase chemistry, several methods for a controlled *in situ* OH production in the laboratory are known. One important method is the pulsed laser photolysis (PLP) technique [45,90], which has been applied in this work. It is based on the fundamental principle of flash photolysis developed by Norrish and Porter [91]. Here, the desired radicals are formed via photodissociation of a precursor molecule. In the special case of PLP the radiation is obtained from a pulsed laser.

In the search for appropriate precursors for kinetic studies in the gas phase, several aspects have to be considered. First, the molecule has to absorb light at a wavelength which is accessible with the provided type of laser. In this work an excimer laser was available, which could be operated at output wavelengths of the common rare gas-fluoride complexes (ArF: 193 nm, KrF: 248 nm, XeF: 351 nm). Subsequently, the precursor should dissociate to the desired radical with a sufficient quantum yield. Here, the co-products as well as by-products from other possible competitive photolysis channels should not disturb the reaction process of interest. In addition, the reaction of the precursor itself with the formed radicals has to be slow, as it is always an unavoidable side reaction. Furthermore, the substance has to exhibit a sufficient vapor pressure, because it has to be evaporated into the gas phase. Moreover, it has to be available with an adequate purity and has to be stable on the time scale of the experiment. Finally, the precursor molecule should not interfere with the detection method with which the reaction of interest is investigated. The method applied in this work is laser-induced fluorescence (LIF) [20,92]. In this case, the radical precursor should not fluoresce significantly at the corresponding wavelengths nor act as an effective quencher for the fluorescence of the OH radicals.

Numerous molecules have already been tested as OH precursor for PLP experiments. All of them exhibit advantageous and disadvantageous aspects, which is why the choice has to be well-considered before starting an experiment. The characteristics of several OH precursors are shortly reviewed in refs. [11,93].

Some precursors found in the literature were excluded from the outset of this work for practical reasons. Nitrous oxide, for example, is photolyzed at a wavelength of 193 nm to form $O(^1D)$ and nitric oxide (see e.g. refs. [94–96]). The $O(^1D)$ radicals then react with a hydrogen source (e.g. water) yielding OH radicals. Although the needed wavelength is achievable with an excimer laser, oxygen is photolyzed and ozone is formed under this condition [27]. As a result, the optical path cannot be conducted through air, which implicates a higher effort on the experimental setup. Moreover, numerous compounds are already photolyzed at this low wavelength. Hence, the photodissociation of the investigated reactant is sometimes a problem. Furthermore, the hydrogen source is an additional substance which has to be present in the reaction mixture. Hence, its side-reaction with OH as well as its fluorescence quenching potential has to be taken into account.

The same reaction mechanism via $O(^1D)$ formation and subsequent reaction with a hydrogen source can be induced by ozone photolysis at wavelengths below 350 nm [27] (see e.g. refs. [97–99]). The problem of the oxygen photolysis in the laser beam path is omitted in this case. However, beside the other above-mentioned disadvantages of this general approach, ozone has to be produced *in situ*. Due this high effort, the application of an ozone/hydrogen source precursor was not considered in this work.

OH generation from the photolysis of ketones, e.g. acetone, in the presence of oxygen is also well known (see e.g. refs. [11, 100, 101]). The formation of OH in this reaction process though is relatively slow, which is why it often superposes the observed reaction. Moreover, oxygen is an effective quencher of OH fluorescence [102].

Another typical OH precursor, which was excluded *a priori*, is nitrous acid. Although it has a large cross section at 351 nm and yields OH with a quantum yield of unity at this wavelength [27], it is mainly used when the studied reactant is nitrous acid itself or only high photolysis wavelengths can be applied (see e.g. refs. [103, 104]). The reason for the rare application can be found in the instability of nitrous acid. Therefore, it has to be produced *in situ* in a relatively complicated synthesis process and can be possibly polluted by nitrogen oxides [105].

An overview of the most important characteristics of the precursors which were considered in this work is given in table 4.1.

A very common OH precursor nowadays is hydrogen peroxide (H_2O_2) (see e.g. refs. [113–115]). It exhibits a reasonable cross section at 248 nm and an OH quantum yield of 2 at this wavelength [27]. Thus, high yields of OH radicals result and no other photodissociation channels nor co-products have to be taken into account. In addition, the reaction of hydrogen peroxide with OH radicals is slow compared to the reaction of other typical OH precursors. However, it exhibits disadvantages, too. First, hydrogen peroxide has a high energy content and can disproportionate explosively in the liquid and gas phase if small amounts of catalytic compounds are present [116]. Thus, it is not available as a pure substance, but only in an aqueous solution. The water, which has a higher vapor pressure than hydrogen peroxide and

precursor	σ(248 nm) [10^{-20} cm^2]	ϕ_{OH}(248 nm)	k_{OH}(298 K) [10^{-12} cm^3 s^{-1}]	P_{vap}(293 K) [mbar]	stable?
H_2O_2	9* [27]	2 [27]	1.8 [27]	2.0 [106]	problematic
t-BuOOH	1.99 [107]	1 [107]	3.56 [107]	5.9 [108]	good
acetylacetone	3056 [109]	unknown	87.8 [110]	10.9 [111]	good
HNO_3	2.00 [27]	~0.9 [79,80]	0.16** [27]	56 [112]	problematic

Table 4.1: Overview of the characteristics of some common OH precursors: σ: absorption cross section at room temperature, ϕ_{OH}: OH quantum yield, k_{OH}: rate coefficient of the reaction precursor + OH, P_{vap}: vapor pressure; *extrapolated on 248 nm; **high-pressure limiting rate coefficient.

does not form an azeotrope with H_2O_2, is typically removed by evacuating the liquid or by bubbling inert gas through. The information about the level of purity which can be achieved with these procedures is vague however. [80, 107, 117, 118] Hence, the presence of water as an impurity and very efficient OH fluorescence quencher [102] is possible even after purification. A second problem can arise from the low vapor pressure of hydrogen peroxide, depending on the method for gas supply which is used. In experiments at pressures below atmospheric pressure the conventional bubbler technique can be applied and sufficient concentrations of hydrogen peroxide in the gas phase can be achieved. However, when mixtures in gas cylinders are prepared higher amounts of the gaseous precursor are needed. Initially, only this technique for gas supply was available for the high-pressure measurements presented herein. Thus, the use of hydrogen peroxide was regarded to be problematic. The high-pressure bubbler which was constructed during this work (cf. subsection 3.1.4) possibly enables the use of hydrogen peroxide as OH precursor, also for high-pressure measurements in the future. However, this alternative was not tested in this work.

The second peroxide which is often applied as OH radical precursor is *tert*-butyl hydroperoxide (*t*-BuOOH) (see e.g. refs. [72, 107]). Its high stability and the absence of any side channels in the photodissociation process are clear advantages. Hence, *t*-BuOOH was tested briefly as an OH precursor in this work. However, a poor signal-to-noise ratio was obtained for the OH-LIF signal at room temperature and pressures above 2 bar. Additionally, the recorded OH lifetime was low resulting from the quite fast reaction of *tert*-butyl hydroperoxide with OH radicals [107].

Another OH precursor, which has already been tested successfully, is acetylacetone (see e.g. refs. [72, 110, 119]). Due to the remarkably high cross section at 248 nm, only small amounts of acetylaceton might be enough for a sufficient OH yield. This can possibly compensate the low vapor pressure and the very high bimolecular rate coefficient of the reaction of acetylaceton with OH. However, the OH quantum yield of the photodissociation is not known yet. As a result, the exact OH yields cannot be predicted and an evaluation of other possible photolysis channels cannot be made. Nevertheless, acetylaceton as OH precursor at pressures above 2 bar was tested in this group [120]. Just like in the case of *t*-BuOOH, an unfavorable signal-to-noise ratio of the LIF signal and a low OH lifetime was observed.

The fourth substance which is named in table 4.1 is nitric acid (HNO_3). Besides its moderate vapor pressure and its very slow reaction with OH, the most prominent advantage of nitric acid as OH precursor is the very favorable signal-to-noise ratio of the OH-LIF signal at pressures above 1 bar. Moreover, the characteristics of its photodissociation, which are discussed in subsection 3.2.4, are well known. Therefore, nitric acid is the preferred OH precursor for high-pressure LIF measurements in this group [65,66]. However, its instability and the formation of nitrous gases as decomposition products has to be considered. According to ref. [70], pure nitric acid can be synthesized easily with the procedure described in subsection 3.1.3. However, in the liquid state the decomposition process can be observed by means of the brownish coloration, which appears on the time scale of hours due to the formation of nitrogen dioxide. While the homogenous decomposition in the gas phase is negligible up to temperatures of 700°C [121], significant amounts of gaseous nitric acid can decompose catalytically on surfaces yielding nitrogen dioxide as well [122]. As the reaction of nitrogen dioxide with OH radicals is very fast [27], the presence of this impurity in the reaction mixture might be a grave error source for kinetic measurements.

With the original approach for gas injection in the PLP/LIF experiment (cf. subsection 3.1.1) the influence of side reactions of the OH precursor and its impurities was not visible. However, different preliminary studies to this work gave rise to the assumption that a systematic error falsifies the experimental results. The determined rate coefficients of the systems HNO_3 + OH [12], DEE + OH [13] and DEE-d10 + OH [14] lay systematically above the values from the literature. The reason of these deviations was assumed to result from the impurities of nitric acid.

However, due to the above named disadvantages of other possible OH precursors and due to the fact that one reactant, which should be studied, was nitric acid itself, no other precursor substance was used. Instead, the impurities of nitric acid were investigated with UV/Vis absorption measurements, which are discussed in this chapter. Thereupon, the setup for gas injection was revised as described in subsection 3.1.2, so that the influence of the side reaction of the OH precursor and its impurities can be quantified and does not cause a systematic error in the determined rate coefficients anymore. However, when the precursor is also the reactant, the precursor and reactant concentration cannot be varied independently. As a result, the precursor/reactant substance must be available on a very high level of purity. Hence, for the study of the system HNO_3 + OH a method was developed which provides pure nitric acid in the gas phase at pressures above 1 bar. The UV/Vis absorption measurements which prove the purity of the gaseous nitric acid supplied with this approach are presented in this chapter, too. Subsequently, the different conclusions which were derived from the UV/Vis measurements were realized in PLP/LIF measurements. In the closing section the consequences for the PLP/LIF approach are discussed and the practically observed advantages of the modifications are illustrated.

4.2 Identification and Quantification of Impurities

According to Stern *et al.* [70] a level of purity of >99.9 % can be achieved with the synthesis approach for nitric acid applied in this work (cf. subsection 3.1.3). The success of the synthesis should be tested in the first UV/Vis measurements. Moreover, the aim was to observe and quantify the amount of impurities which is possibly formed in the subsequent thawing and evaporation process.

4.2.1 Experimental Procedure

The flask containing the freshly prepared frozen nitric acid was installed in the static UV/Vis setup, which is described in subsection 3.3.1. The solid substance was kept under cooling with liquid nitrogen during the preparation of the experiment. Before starting the measurements, the whole setup was evacuated and the pressure rise after disconnecting the pump was examined with a pressure transducer. When the tightness of the apparatus was ensured the zero baseline was recorded with the UV/Vis spectrometer. Then, the connection valve between vacuum pump and reservoir flask was closed and the cooling of the flask was removed. Altogether 36 UV/Vis spectra were recorded during the thawing and evaporation process of the nitric acid. At the beginning, a measurement was conducted every one to two minutes. At the end of the experiment the time intervals between the measurements were increased. The experiment was stopped after approx. 100 min.

All spectra were recorded between 200 and 800 nm in intervals of 1 nm. The averaging time was tuned to 0.1 s, while the spectral band width amounted to 2 nm (fwhm). The time stamps of the output files obtained from the spectrometer were taken as time specification. The ambient temperature amounted to approximately 295 K.

The following chemicals were used: H_2SO_4 98 %, Roth; KNO_3 $\geq$ 99 %, Roth; He > 99.999 %, Air Liquide.

4.2.2 Results and Discussion

A selection of spectra recorded at different times after removing the cooling is shown in figure 4.1. Below 300 nm a flank of a diffuse band, which increases rapidly with time, can be observed. It was assigned to nitric acid. Some small deviations in the shape of the spectrum in comparison to the literature data [27] may arise from the initially low temperature and the interactions between the molecules at later times due to the high

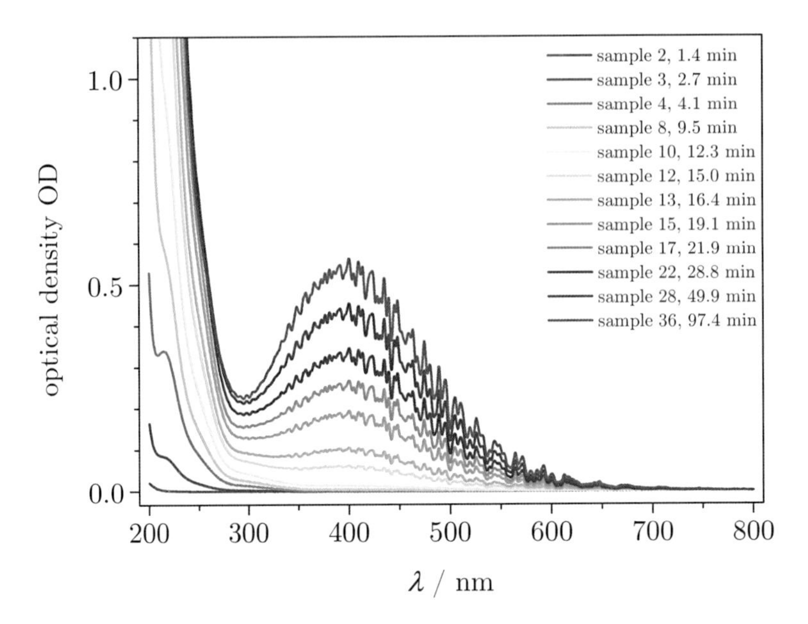

Figure 4.1: Absorption spectra of vapor over freshly synthesized HNO_3 during the thawing process at different times after removing the cooling.

concentration. Nevertheless, the rise in the HNO_3 vapor concentration can be observed qualitatively. From sample 22 at 28.8 min no significant change in this band was obtained anymore, which is why it was assumed that the vapor pressure of nitric acid is reached at that point.

Additionally, a structured band with a maximum at approx. 400 nm was obtained. It is absent in the first spectra. However, from sample 10 at 12.3 min it increases, even when the partial pressure of nitric acid has reached its vapor pressure. This band could clearly be assigned to nitrogen dioxide (NO_2). The good agreement between the measurement and the normalized spectrum from the literature [123] is illustrated in figure 4.2. The deviations in the maxima and minima of the fine structure result from the differences in the spectral resolution. While in the measurements from ref. [123] it ranged from 0.24 to 0.48 nm, the spectral resolution in the experiments of this work amounted to 2 nm (fwhm).

The concentration of NO_2 was determined from the optical density at 400 nm OD(400 nm) and the optical path length d with equation 2.10 according to

$$[NO_2] = \frac{2.303\ \mathrm{OD}(400\ \mathrm{nm})}{\sigma(400\ \mathrm{nm})d}\,. \tag{4.1}$$

For the absorption cross section of NO_2 $\sigma(400\ \mathrm{nm})$, the values at 399.93 and 400.12 nm of Bogumil *et al.* [123] at 293 K were averaged, which yielded $\sigma(400\ \mathrm{nm}) = 6.709 \cdot 10^{-19}\ \mathrm{cm}^2$. Although the evaluation this way includes a rather large error due to the above mentioned

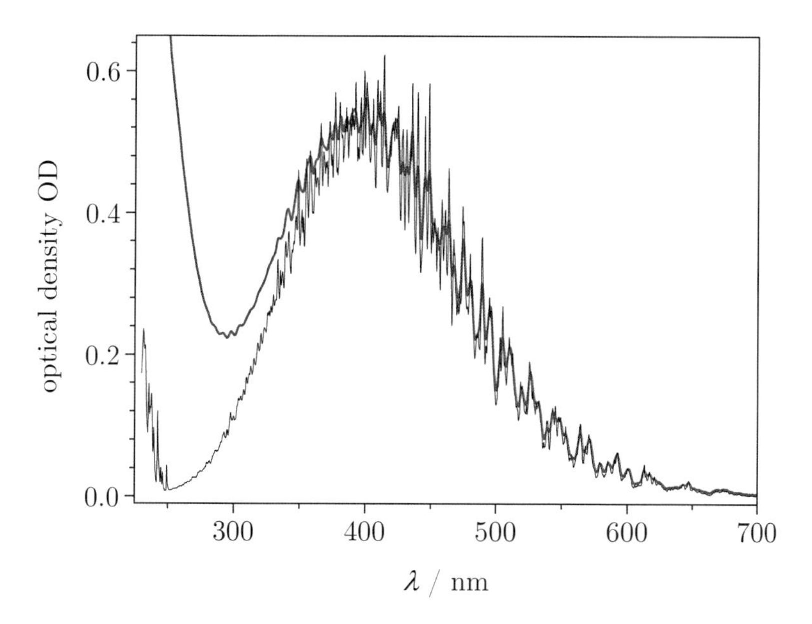

Figure 4.2: Comparison between the recorded spectrum (sample 36) at 97.4 min after removing the cooling (—) and the spectrum from ref. [123] (—). The literature spectrum was normalized by means of the NO_2 concentration calculated from the optical density at 400 nm (see text).

differences in the spectral resolutions, it suffices for an estimation of the nitrogen dioxide concentration. An accurate determination of $[NO_2]$ was not necessary in these measurements.

The temporal evolution of the resulting $[NO_2]$ is depicted in figure 4.3. It illustrates that nitric acid on a high level of purity is obtained from the synthesis as described in subsection 3.1.3. However, during the thawing and evaporation of the initially frozen nitric acid, decomposition already takes place and a fast increase of the NO_2 concentration can be observed. When the vapor pressure of nitric acid is reached, the decomposition decelerates, but the NO_2 concentration still increases significantly. These results can be explained in terms of the decomposition of nitric acid in the liquid phase and catalytically in the gas phase on the available surfaces [70, 122, 124].

4.2.3 Conclusions

The experimental findings described in this section lead to the conclusion that significant amounts of NO_2 are present when nitric acid is evaporated into the gas cylinders during mixture preparation, as described in subsection 3.1.3. However, a quantitative prediction of the initial NO_2 content of the gas mixtures is not possible, as the heterogeneous decomposition of gaseous nitric acid is dependent on the kind, and size of the available surface [122].

Nevertheless, the reaction of NO_2 with OH radicals, with a rate coefficient of approx.

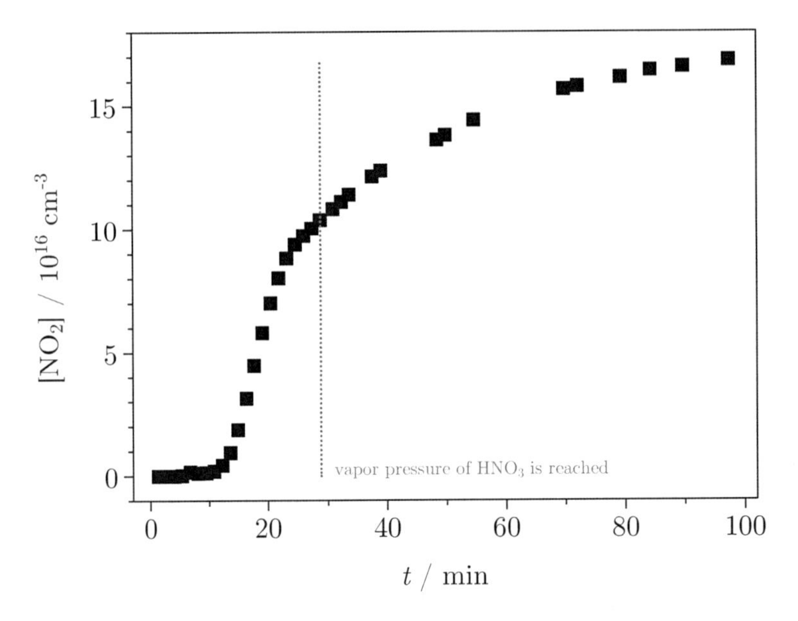

Figure 4.3: Temporal evolution of the NO_2 concentration in the gas phase over freshly synthesized nitric acid during the thawing and evaporation process.

$1 \cdot 10^{-11}$ cm^3 s^{-1} at 300 K [27], is fast. It exhibits a positive pressure and a negative temperature dependence under the conditions relevant for this work [27]. Consequently, small amounts of the impurity can already bias the kinetic measurements. A general neglect of the NO_2 + OH side reaction thus is not legitimate.

4.3 Purity of Nitric Acid Gas Mixtures

For the examination of the amount of nitrogen dioxide in gas mixtures containing nitric acid, UV/Vis measurements were carried out using HNO_3/helium gas mixtures as typically applied in the usual PLP/LIF experiment. The temporal profile of the NO_2 concentration was studied and the amount of decomposed HNO_3 was determined. Necessary modifications of the PLP/LIF setup were derived from these measurements.

4.3.1 Experimental Procedure

A mixture with 5.29 mbar of freshly synthesized nitric acid and 30.43 bar helium was prepared in an electropolished stainless steel test gas cylinder as described in subsection 3.1.3. The measurements were carried out with the slow-flow UV/Vis absorption setup, which is described in section 3.3. The original approach for gas injection was applied (cf. subsection 3.1.1).

In preliminary measurements it turned out that, at low gas flows, the measured NO_2 concentration is dependent on the gas flow. It was assumed that the retention time is too high, so that significant amounts of nitric acid decompose inside the apparatus along the way. Above 3 slm $[NO_2]$ was independent from the flow rate. Hence, a flow of 4 slm was chosen for the measurements presented here.

Moreover, a high residual content of nitric acid in the setup was detected after the measurements, even when the apparatus was purged with inert gas for several hours with a high gas flow. When the purging was stopped by setting the gas flow to zero the HNO_3 and NO_2 bands increased rapidly in the absorption spectra. However, with a slow flow of inert gas the bands disappeared completely. As a result, the baselines of the spectra described here were recorded under a constant flow of pure helium, which amounted to 1 slm.

14 series of measurements on different days were recorded. Altogether, the NO_2 content in the nitric acid gas mixture was monitored over 45 days. In one series of measurements, first the zero baseline was recorded. Then, its stability was checked in a subsequent measurement of an additional zero spectrum. When the baseline stability was ensured the helium cylinder was exchanged by the nitric acid mixture and the gas flow was regulated. After purging the apparatus for approx. 1 min, the absorption measurement was started. On the first day six spectra were recorded, while only one spectrum was measured in the following series of measurements. During the measurements, the pressure of the gas cylinder altogether decreased from 29.7 to 22.7 bar.

The wavelength ranged from 200 to 800 nm in all records. The step size amounted to 0.5 nm. An averaging time of 0.1 s and a spectral band width of 2 nm (fwhm) were chosen. The point in time of the record was specified with the time stamps of the output files of the spectrometer. The measurements were conducted and the mixture was stored at room temperature.

The chemicals which were used were the following: H_2SO_4 98 %, Roth; $KNO_3 \geq 99$ %, Roth; He > 99.999 %, Air Liquide.

4.3.2 Results and Discussion

The concentration of NO_2 was calculated analogously to the analysis described in section 4.2. For the determination of the HNO_3 concentration the optical density at 202 nm was used, corrected by the partial optical density from the absorption of NO_2 at 202 nm. Thus, $[HNO_3]$ was obtained on the basis of equation 2.10 by the expression:

$$[\mathrm{HNO_3}] = \frac{2.303\ \mathrm{OD}(202\ \mathrm{nm}) - \sigma_{\mathrm{NO2}}(202\ \mathrm{nm})[\mathrm{NO_2}]d}{\sigma_{\mathrm{HNO3}}(202\ \mathrm{nm})d} \, . \tag{4.2}$$

The absorption cross section of HNO_3 at 202 nm and 298 K was taken from [83], while for NO_2, an average value between 200 and 205 nm of the cross sections reported in ref. [125]

was applied.

The measured absorption spectra were compared to the sum of the literature spectra of HNO_3 from ref. [83] and of NO_2 from ref. [123] normalized by the calculated concentrations. Figure 4.4 shows the result for the measurement carried out after 3.9 days. The very good agreement, which can be observed in figure 4.4, was obtained for all spectra.

An initial HNO_3 concentration $[HNO_3]_0$ was defined by the amount of HNO_3 which would be present in the mixture without decomposition. It was calculated as the sum of $[HNO_3]$ and $[NO_2]$ obtained from the measured spectra as well as from the filling pressures of the gas mixture and the current pressure in the setup under the assumption of ideal gases. After the first day, the resulting $[HNO_3]_0$ agreed well within 4 %. However, on the first day the values obtained from the pressures overestimated the determination from the spectra by a maximum of 33 % for the first measurement. The deviation decreased to 11 % after 3 hours. This observation was explained with the incomplete homogenization of the gas mixture directly after preparation.

Hence, to evaluate the temporal evolution of the NO_2 impurity in nitric acid gas mixtures, the $[HNO_3]_0$ calculated from the spectra was used. A plot of the ratio $[NO_2]/[HNO_3]_0$ against the time is shown in figure 4.5.

It can be seen that already 5 % of the apparent HNO_3 filling pressure during mixture preparation results from NO_2. In the following, a very fast increase in the $[NO_2]/[HNO_3]_0$ ratio to approx. 12 % is observed at the first day. The decomposition decelerates in the following to a change of around 0.9 percentage points per day between 3 and 20 days. After 45 days approx. 40 % of the filled in nitric acid is decomposed to NO_2.

These results are not quantitatively transferable on other HNO_3 mixtures due to the high dependency of the HNO_3 decomposition on the nature of the available surfaces (cf. section 4.2). Nevertheless, it is proven with these measurements that high amounts of HNO_3 can decompose during the storage of HNO_3 mixtures.

4.3.3 Conclusions and Revision of the Gas Injection Method

With the application of the original approach for gas injection in the PLP/LIF experiment, a significant change of the reactant concentration was only possible by draining off the gas mixture partly. As this is a very time and resource consuming approach, the usual measurements at one fixed temperature were made in a narrow range of the reactant concentration. The bimolecular rate coefficients were obtained by dividing the pseudo-first order rate coefficients by the reactant concentration. Thus, side reactions had to be negligible or their pseudo-first order rate coefficients had to be known and abstracted before.

Only at few temperatures the reactant concentration was varied over a wide range. With these measurements, the linearity between the reactant concentration and the pseudo-first order rate coefficients was checked. However, a simultaneous change in the pressure could

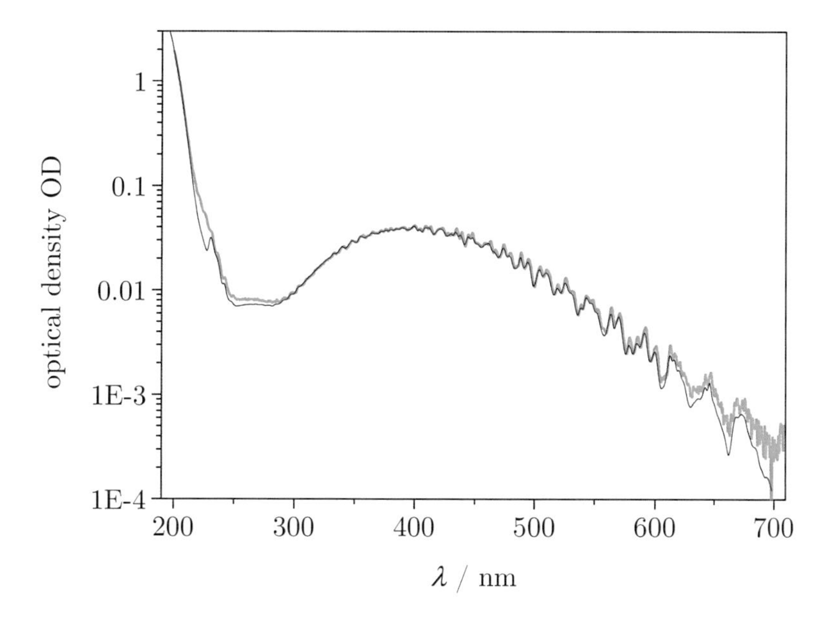

Figure 4.4: Comparison between the spectrum of the HNO_3 mixture ($P = 27.0$ bar) recorded 3.9 days after preparing the mixture (—) and the calculated cumulative spectrum based on the absorption cross sections from refs. [27,123] (—) normalized by means of the HNO_3 and NO_2 concentrations (see text).

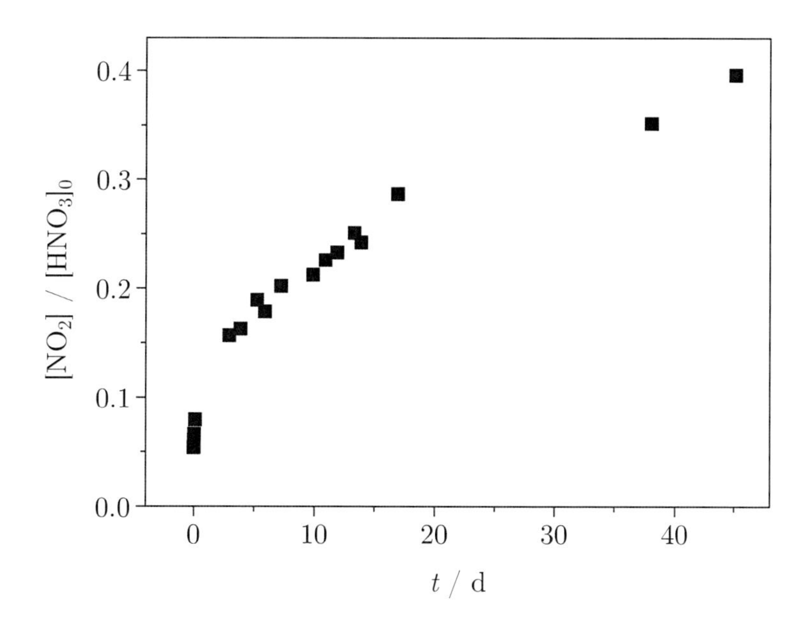

Figure 4.5: Temporal evolution of the ratio $[NO_2]/[HNO_3]_0$ in the HNO_3 mixture.

not be avoided. Thus, such a data evaluation was only possible for pressure independent reactions. Moreover, the OH precursor concentration, as well as the concentration of its impurities, decreased proportionally to the reactant concentration decline. Hence, the influence of the side reactions on the measured overall pseudo-first order rate coefficient could not be seen from these pseudo-first order plots and the intercept of the linear fit scattered around zero.

However, the studies presented in this section and in section 4.2 suggest that the NO_2 + OH side reaction in kinetic measurements, with the use of HNO_3 as OH precursor and gas mixtures for gas supply, cannot be generally neglected.

Hence, the approach for gas injection was revised in the following. The revised setup is described in subsection 3.1.2. With the three different gas supply lines the concentrations of the different components in the reaction mixture can be varied flexibly and independently. Consequently, the reactant concentration can be altered easily at a fixed OH precursor concentration and overall pressure. As a result, all bimolecular rate coefficients can be obtained from a plot of the pseudo-first order rate coefficients against the reactant concentrations which results from one series of measurements at a defined temperature and pressure. Moreover, the rates of the side reactions of the OH precursor and its impurities with OH radicals stay constant during such a series of measurements. As a result, their pseudo-first order rate coefficients appear as the intercept in the pseudo-first order plots. Thus, the influence of these side reactions on the overall measured rate coefficient can directly be seen and they do not cause a systematic error in the bimolecular rate coefficient of interest.

The co-products of the heterogeneous decomposition of $[HNO_3]_0$ are water and oxygen [122]. Both are known as very effective quenchers for OH fluorescence [102]. Hence, the decomposition of HNO_3 inside the gas cylinders not only causes an increase in the rate of the side reactions but also a decrease of the signal-to-noise ratio. According to the findings presented in this section, it is therefore recommended that the age of the HNO_3 gas mixtures is kept low.

4.4 Supply of Pure Nitric Acid in the Gas Phase

To enable the investigation of the kinetics of the reaction HNO_3 + OH at high pressures, a gas supply method was needed which provides nitric acid in the gas phase on a high level of purity. In the work of Brown *et al.* [10] it has already been reported that pure nitric acid under slow-flow conditions can be obtained with the bubbler technique at pressures below atmospheric pressure. Hence, a bubbler cylinder suitable for high pressures was developed, to make the application of this technique possible also for measurements above 1 bar. The apparatus is described in subsection 3.1.4.

To investigate the level of purity of the gaseous nitric acid obtained with this technique, and to prove its suitability for the planned kinetic study, UV/Vis absorption measurements were conducted preliminary to the PLP/LIF experiments on the kinetics of the system HNO_3 + OH.

4.4.1 Experimental Procedure

The measurements presented in this section were carried out with the slow-flow UV/Vis absorption setup (cf. section 3.3). The revised setup for gas injection, which is described in subsection 3.1.2, was applied using one gas supply unit. The helium line was only used for purging the setup before and after the measurements. For the pressure regulation in the cell and the bubbler, no pressure controllers were available at that time. Hence, the bubbler pressure was controlled with the pressure reducer of the helium cylinder. The pressure in the absorption cell was regulated with a needle valve. All experiments were carried out at room temperature.

Before starting the experiment, the pressure in the absorption cell was set to the desired value and the whole setup was purged with pure helium for approx. 1 min. Subsequently, the baseline was recorded under a flow of helium. When the bubbler was used for gas supply, the pressure inside the bubbler was regulated and bubbling was started. Then, the absorption spectra were recorded.

In general, two different types of absorption measurements were carried out. Rough spectra were measured in a wavelength range between 200 and 800 nm with a step size of 1 nm and an average time of 0.1 s. In fine measurements the wavelength ranged from 432.5 to 450.0 nm. Here, wavelength steps of 0.2 nm and an average time of 40 s was chosen. Thus, the record of a spectrum in the rough measurements took one minute, while for the record of a fine spectrum one hour was needed. The spectral band width amounted to 0.52 nm in all measurements.

The rough measurements were carried out for a simultaneous observation of $[HNO_3]$ and $[NO_2]$ when the bubbler was used for gas supply. It was filled with 100 % red fuming nitric acid, 100 % freshly prepared nitric acid (according to the procedure described in subsection 3.1.3) or a ternary mixture of approx. 50 wt% nitric acid, 28 wt% water and 22 wt% sulfuric acid. Absorption spectra were recorded at different time delays after starting the bubbling and at different bath gas flows. The time stamps of the output files of the spectrometer were used for time specification.

The fine measurements served for the examination of very low NO_2 concentrations. Thus, the limit of detection of $[NO_2]$ was determined at first. For this purpose, 59 blank spectra on three different days were recorded in fine measurements. The measurements were carried out under a slow helium flow of 0.5 slm or in air with the absorption cell disconnected from the gas injection setup.

To prove the sensitivity for the quantitative detection of $[NO_2]$, a fine measurement was carried out with an NO_2 gas mixture in a test gas cylinder with helium as bath gas. Low NO_2 concentrations of approx. 4 ppm were achieved with a dilution series. After allowing homogenization of the mixture for at least 12 hours, the gas cylinder was connected to the setup as gas supply unit. The spectrum was recorded with a gas flow of 1 slm.

Subsequently, fine measurements were carried out in order to determine the stationary NO_2 concentration which is reached after some time of bubbling with the above-mentioned ternary mixture. Here, the gas flow was set to 5 or 10 slm.

The ternary mixture was prepared with 82 vol% of 65 wt% nitric acid and 18 vol% of 98 wt% sulfuric acid. Generally, the following chemicals were used: HNO_3 (red fuming) 100 %, Sigma-Aldrich; HNO_3 65 %, Fluka; H_2SO_4 98 %, Roth; KNO_3 $\geq$ 99 %, Roth; NO_2 99.5 %, Air Liquide; He $>$ 99.999 %, Air Liquide.

4.4.2 Results with Concentrated Nitric Acid

In a first series of rough measurements the change of the absorption spectrum of the emanating vapor was observed when helium was bubbled through red fuming nitric acid. Here, a gas flow of 5 slm was chosen, while the pressure in the absorption cell was set to approx. 5 bar. The pressure inside the bubbler ranged from 23.6 to 28.7 bar. Selected spectra at different times after starting the bubbling are shown in figure 4.6.

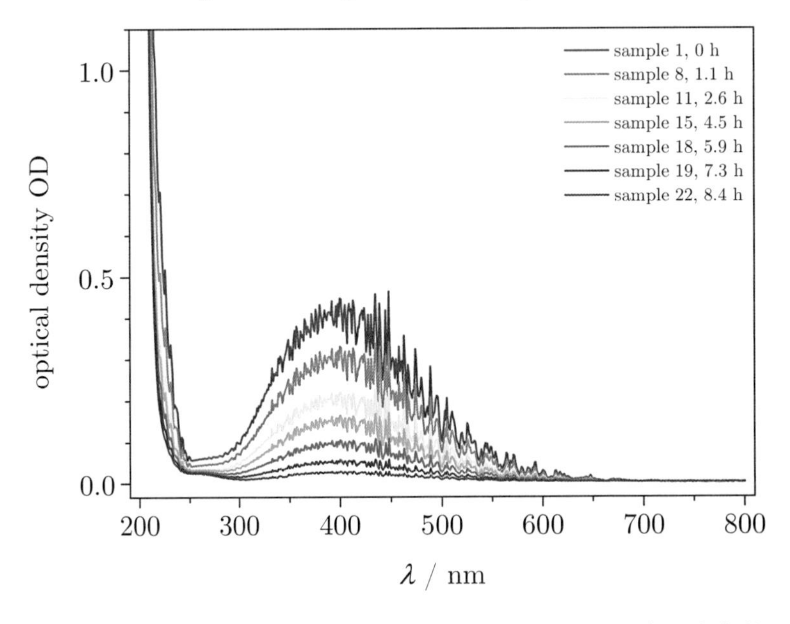

Figure 4.6: Absorption spectra of the vapor of red fuming nitric acid with helium as bath gas at different times after starting the bubbling.

A clear reduction of the NO_2 concentration can be observed by means of the broad absorption band around 400 nm. However, even after 8.4 hours of bubbling the NO_2 concentration was still too high for the requirements of this work. As a high amount of the liquid had already been evaporated, so that not much was left inside the bubbler, the application of red fuming nitric acid in the further studies was excluded.

The use of freshly prepared nitric acid was considered, too. The flow rate was varied between 1 and 5 slm. The pressure inside the absorption cell ranged from 5 to 21 bar, while the bubbler pressure amounted to 23 to 29 bar. Although the initial NO_2 content could be reduced significantly in comparison to red fuming HNO_3, long bubbling times for purification were also needed and the amount of the residual NO_2 could not be reproduced clearly. Hence, the use of freshly prepared nitric acid for the subsequent study of the reaction HNO_3 + OH is regarded to be problematic, as long as a simultaneous detection of the NO_2 concentration during the PLP/LIF measurements is not possible.

4.4.3 Results with a Ternary Mixture

Due to the problems in the use of red fuming and freshly prepared nitric acid, the suitability of the ternary mixture was tested. Generally, the pressure inside the absorption cell was varied between 3.4 and 38.4 bar, while the pressure inside the bubbler ranged from 20.9 to 41.0 bar.

In the rough spectra of the vapor of a freshly prepared ternary mixture no NO_2 was observed at different gas flows between 1 and 5 slm even directly after starting the bubbling. Hence, the concentration of HNO_3 was determined on the basis of equation 2.10 directly from the optical density at 208 nm according to:

$$[\mathrm{HNO_3}] = \frac{2.303\ \mathrm{OD}(208\ \mathrm{nm})}{\sigma(208\ \mathrm{nm})d} . \tag{4.3}$$

The absorption cross section at 208 nm was taken from ref. [83].

By means of these concentrations the absorption cross sections over the whole wavelength range were calculated from the measured spectra. A very good agreement between the measurements and the literature data was obtained. It is illustrated in figure 4.7.

To be able to prove the purity of HNO_3 on the level which is needed for the PLP/LIF measurements on the system HNO_3 + OH a special evaluation procedure for the subsequent fine measurements was worked out. The measured spectra were fitted to the absorption cross sections from ref. [123] in a least-squares fit according to the following equation:

$$\sigma(\lambda) = a + b\lambda + \frac{\mathrm{OD}(\lambda)}{[\mathrm{NO_2}]d} . \tag{4.4}$$

In this fit a sloping offset was assumed. This was necessary due to a slight temporal baseline

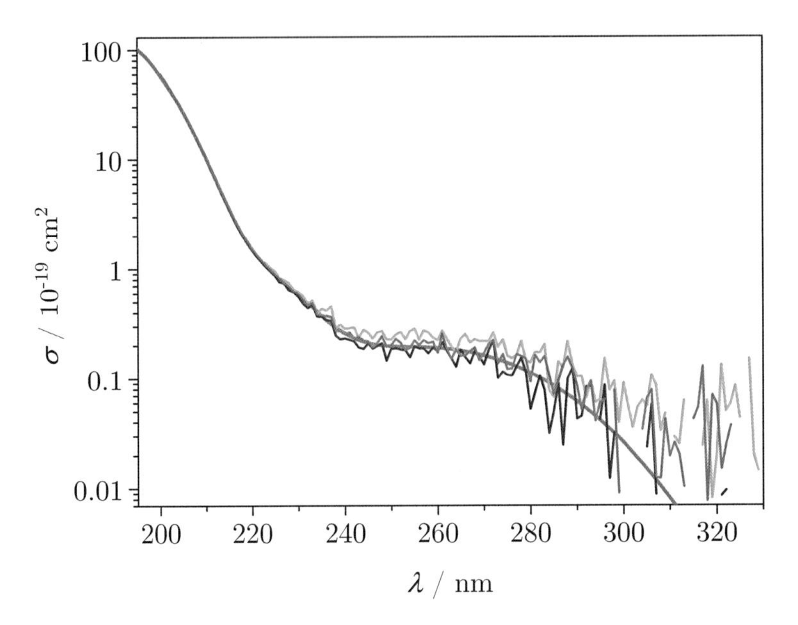

Figure 4.7: Absorption cross sections of gaseous HNO_3: this work, $f = 1$ slm, $[HNO_3] = 3.62 \cdot 10^{16}$ cm^{-3} (—); this work, $f = 3$ slm, $[HNO_3] = 3.69 \cdot 10^{16}$ cm^{-3} (—); this work, $f = 5$ slm, $[HNO_3] = 2.81 \cdot 10^{16}$ cm^{-3} (—); Burkholder *et al.* [83] (—). The absorption cross sections in this work were obtained by normalizing the measured optical density to the literature value of σ(208 nm) from ref. [83].

drift, which was observed in all UV/Vis measurements. Due to the long acquisition time in the fine measurements, this led to a significant increase or decrease in the background signal during one recording. In good approximation, the variation was linear in time. The fitting was carried out with a program written by Pfeifle in Fortran90 [126]. With this procedure the needed sensitivity of the NO_2 detection could be achieved.

In general, the limit of detection (LOD) is defined by [127]:

$$\mathrm{LOD} = \overline{x_{\mathrm{bl}}} + 3\, s(\overline{x_{\mathrm{bl}}})\,, \tag{4.5}$$

where $\overline{x_{\mathrm{b}}}$ is the mean target value and $s(\overline{x_{\mathrm{b}}})$ its standard deviation obtained from an adequate number of blank measurements. 59 of such blank measurements were carried out for the determination of the limit of detection for the NO_2 concentration $\mathrm{LOD}_{\mathrm{NO2}}$. The mean blank value $[NO_2]_{\mathrm{bl}}$ and the standard deviation $s([NO_2]_{\mathrm{bl}})$ were obtained with the fitting procedure described above. The following value for the limit of detection for $[NO_2]$ resulted:

$$\mathrm{LOD}_{\mathrm{NO2}} = 3.2 \cdot 10^{13}\ \mathrm{cm}^{-3}\,. \tag{4.6}$$

Subsequently, fine spectra of the NO_2 mixture in the test gas cylinder and the vapor

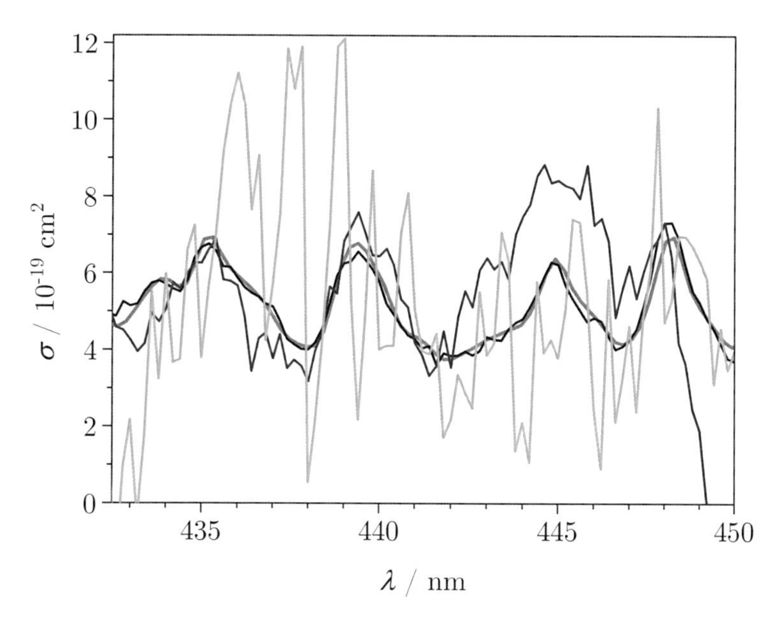

Figure 4.8: Absorption cross sections of NO_2: this work, NO_2 mixture, $[NO_2] = 3.6 \cdot 10^{14}$ cm^{-3} (—); this work, bubbled ternary mixture, $f = 5$ slm, $[NO_2] = 9.2 \cdot 10^{13}$ cm^{-3}(—); this work, bubbled ternary mixture, $f = 10$ slm, $[NO_2]$<LOD_{NO2} (—); Burkholder *et al.* [83] (—). All absorption cross sections from this work were obtained by a least squares fit according to equation 4.4 to the literature spectrum from ref. [83].

emanating from the ternary mixture in the bubbler were recorded. Figure 4.8 shows examples of fitted absorption cross sections from these measurements under different conditions compared to the literature spectrum.

A good agreement with the literature data is observed in the spectrum from the NO_2 mixture, although a quite low NO_2 concentration of $[NO_2] = 3.6 \cdot 10^{14}$ cm^{-3} was detected. A significant decrease in the signal-to-noise ratio is observed for the measurements with the ternary mixture at a gas flow of 5 slm. However, the peaks in the spectrum can still be reproduced and the obtained NO_2 concentration of $9.2 \cdot 10^{13}$ cm^{-3} is above the detection limit. In contrast, only a random noise was obtained from the measurements with the ternary mixture at a gas flow of 10 slm. Hence, it is plausible that the calculated value for $[NO_2]$ in this case is below the detection limit.

The concentrations of HNO_3 in these measurements were estimated. In the preliminary rough measurements $[HNO_3]$ was determined with the help of equation 4.3. From this value, the HNO_3 concentration in the vapor over the ternary mixture $[HNO_3]_{vap}$ was calculated under consideration of the ratio between the current pressure inside the bubbler P_b and the current pressure in the absorption cell P_c. Although this vapor concentration changes slowly with time, it could be assumed to be constant during one series of measurements in good approximation. Consequently, the concentration of HNO_3 in the fine measurements was determined from the current pressures according to

$$[HNO_3] = \frac{P_c}{P_b}[HNO_3]_{vap} \,. \tag{4.7}$$

The concentration of HNO_3 in both of the measurements with the ternary mixture, which are shown in figure 4.8, was approx. $2{\cdot}10^{17}$ cm^{-3}. Hence, the ratio $[NO_2]/[HNO_3]$ amounted to $3.9{\cdot}10^{-4}$ for the measurement with 5 slm and to less than $1.6{\cdot}10^{-4}$ for the measurement with 10 slm. Generally, in all fine measurements with a gas flow of 10 slm, $[NO_2]$ was below the detection limit. This observation was independent of the pressure inside the bubbler, which was varied between 20.9 and 41.0 bar in these measurements. With a gas flow of 5 slm the concentration of NO_2 varied slightly from measurement to measurement near the detection limit. The above-mentioned ratio $[NO_2]/[HNO_3]$ can be seen as an estimation of the upper limit which can be achieved at this flow rate.

From these values the influence of the error caused by the NO_2 + OH side reaction on the measurement of the rate coefficient of the reaction HNO_3 + OH (reaction R1) can be estimated. It can be characterized by the ratio of the corresponding pseudo-first order rate coefficients according to:

$$\frac{k'_{NO2}}{k'_1} = \frac{k_{NO2}}{k_1}[NO_2]/[HNO_3] \,. \tag{4.8}$$

Here, k'_{NO2} is the pseudo-first order rate coefficient of NO_2 + OH and k_{NO2} the corresponding bimolecular rate coefficient.

As k_{NO2} still increases significantly with the pressure above 1 bar and shows a pronounced negative temperature dependence, the estimation was made for a pressure of 10 bar and a temperature of 300 K. For k_{NO2} the recommendation from ref. [128] was taken, while for k_1 the proposed high-pressure limit of Brown *et al.* [10] was applied.

Values of 6.5 % for 5 slm and <2.7 % for 10 slm were obtained for the upper limit of k'_{NO2}/k'_1. As a result, the level of purity with respect to NO_2 which can be obtained with the high-pressure bubbler technique described herein is generally sufficient for the kinetic measurements on the system HNO_3 + OH.

In some of the measurements the PLP/LIF reaction cell for low temperatures and high pressures (cf. subsection 3.2.2) was mounted between the bubbler and the absorption cell. No change in the detected NO_2 concentrations arose. Thus, it can be excluded that significant amounts of HNO_3 decompose in the capillaries or in the PLP/LIF cell at the chosen flow rates.

When the ternary mixture was stored over night or longer inside the bubbler high amounts of NO_2 could be detected in measurements directly after starting the bubbling. However, the impurity disappeared within a few minutes when helium was bubbled through with a flow rate of 10 slm.

A remaining issue in the application of a ternary mixture, like it has been applied in this study, is the water content of the emanating vapor from the bubbler. Although it can be assumed that a significant reduction can be achieved by the addition of the sulfuric acid to the concentrated nitric acid [129, 130], a residual humidity is probable [131]. Nevertheless, it is shown by Carl *et al.* [132] that even the presence of high amounts of water does not influence the kinetics of reaction HNO_3 + OH. Consequently, for the study of this system the above-named ternary mixture is suitable for the supply of nitric acid in the gas phase.

4.4.4 Conclusions

In the study presented in this section a method for the supply of nitric acid in the gas phase under slow-flow conditions and pressures above atmospheric pressure on a high level of purity was tested. In the application of red fuming, and freshly synthesized nitric acid difficulties arose to achieve the high level of purity which is needed for the measurements of k_1. Moreover, the residual NO_2 concentrations were not clearly reproducible and thus a permanent control of the NO_2 concentration would be necessary in the application of these substances. Consequently, the use of red fuming and freshly synthesized pure nitric acid is not recommendable for the PLP/LIF measurements of k_1 with the present setup.

The use of a ternary mixture with 50 wt% nitric acid, 28 wt% water and 22 wt% sulfuric acid in the high-pressure bubbler, however, is suitable for this purpose. The NO_2 content after a fresh preparation of the mixture is negligible so that reliable measurements of k_1 become possible with this approach. Generally, the $[NO_2]/[HNO_3]$ ratio in the gas phase increased significantly when the ternary mixture was stored inside the bubbler for more than a few hours without bubbling. However, the formed NO_2 could be removed quickly by bubbling inert gas though the liquid. Hence, a purification of the ternary mixture in this way for 20 min with a gas flow of 10 slm is recommended after storage.

The stationary residual $[NO_2]$ was dependent on the gas flow. For a flow rate of 5 slm a ratio $[NO_2]/[HNO_3]$ of maximum $3.9 \cdot 10^{-4}$ was obtained, while the NO_2 concentration with a flow rate of 10 slm was always below the detection limit. So $[NO_2]/[HNO_3]$ is always less than $1.6 \cdot 10^{-4}$ in this case. The upper limit of the ratio between the pseudo-first order rate coefficients of the reactions NO_2 + OH and HNO_3 + OH was estimated to be 6.7 % for 5 slm and <2.7 % for 10 slm at 300 K and 10 bar. For lower temperatures and pressures it is expected to decrease significantly.

Hence, it can be concluded that an error caused by the NO_2 impurity can be neglected when a purified ternary mixture in the high-pressure bubbler is used for gas supply and the flow rate of the bubbling is higher than 5 slm.

The application of the bubbler technique in a slow-flow setup requires certain features. Especially a flow controlling unit between bubbler and actual setup is reasonable so that a control of the emanating concentration is possible without the need of changing the vapor pressure. Moreover, a further dilution can be achieved with this approach by adding bath gas through an additional gas line behind the bubbling unit. All these features could be provided with the revised setup for gas injection, which was established to handle the impurities of HNO_3 in gas mixtures (cf. section 4.3) and is described in subsection 3.1.2.

4.5 Consequences for the PLP/LIF Experiment

The results of the study on the purity of HNO_3 led to the revision of the gas injection setup for the PLP/LIF experiments. With the revised approach, the use of the high-pressure bubbler is possible and the problem of the NO_2 impurity in the gas mixtures containing HNO_3 is overcome. Moreover, reliable measurements at low reactant concentrations for the study of very fast OH reactions are enabled with the modifications. All of these improvements are illustrated in this section by discussing the PLP/LIF results obtained with the original and revised approach.

4.5.1 Measurements with the High-Pressure Bubbler

One reason for the modification of the gas injection setup was to enable the use of the high-pressure bubbler especially to obtain gaseous nitric acid on a high level of purity. In first test measurements on the kinetics of the reaction

$$HNO_3 + OH \xrightarrow{k_1} \text{products}, \qquad \text{(R1)}$$

the suitability of the new gas supply and gas injection setup was examined. The experimental study on this system was then continued by Zügner [76]. Moreover, a theoretical study on this system was carried out by Pfeifle [133]. As the experimental investigation has not been finished yet, only the results of the first test experiments, which were carried out in this work, are briefly presented here.

19 single measurements were conducted at 296 K, 10 bar and different HNO_3 concentrations. The repetition rate of the experiment amounted to 10 Hz. The concentration of HNO_3 was varied by changing the pressure inside the bubbler, which ranged from 13.1 to 80.8 bar. The flow rate was regulated with the flow controller mounted after the bubbler and amounted to 10 slm in every measurement. The helium line was not used for a further dilution.

The HNO_3 concentration was determined by the help of the absorption unit (cf. subsection 3.2.3) operated with a Zn-lamp at 214 nm and the absorption cross section from ref. [83] at 293 K. It ranged from $2.6 \cdot 10^{16}$ to $11.5 \cdot 10^{16}$ cm^{-3}, while from the estimation of [OH] according to equations 3.2 and 3.1 values between $1.0 \cdot 10^{12}$ and $4.3 \cdot 10^{12}$ cm^{-3} were obtained.

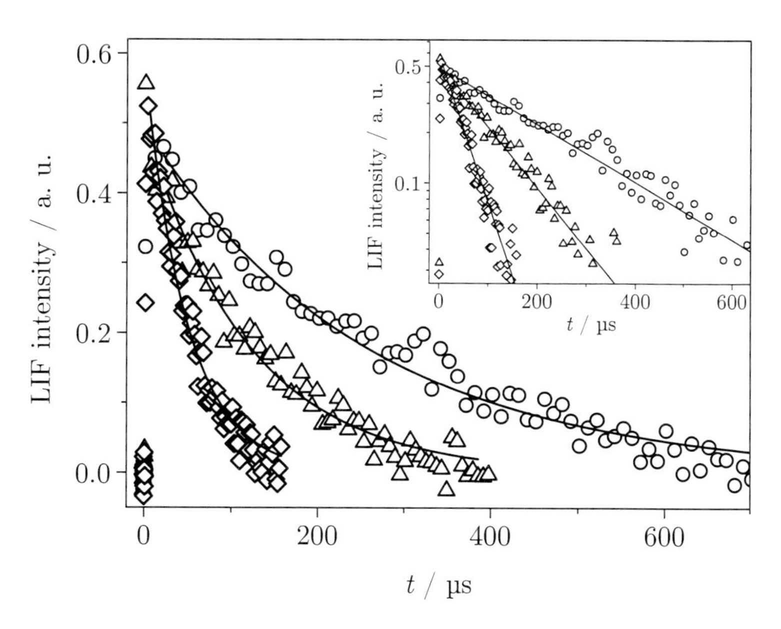

Figure 4.9: LIF intensity-time profiles of the reaction HNO_3 + OH at 10 bar, 296 K, and different [HNO_3]: $1.15 \cdot 10^{17}$ cm^{-3} (○), $4.90 \cdot 10^{16}$ cm^{-3} (△), $2.59 \cdot 10^{16}$ cm^{-3} (◇). For a better illustration the linearized signals are shown as well.

For each measuring point at a given time delay between photolysis and detection 10 single values were averaged.

Representative LIF intensity-time signals are depicted in figure 4.9. All measured signals could be fitted adequately with a monoexponential decay function according to

$$I = I_0 \exp\left(-k_1' t\right) \ . \tag{4.9}$$

An overview of the so-obtained pseudo-first order rate coefficients k_1' and the corresponding measuring conditions can be found in appendix A.1.

All k_1' were plotted against the HNO_3 concentration, which is shown in figure 4.10. A linear relation with a negligible intercept resulted. The bimolecular rate coefficient k_1 was determined from the slope of the best linear fit. A value of $(1.6 \pm 0.2)\ 10^{-13}$ cm^3 s^{-1} was obtained.

In figure 4.11 k_1 at room temperature is plotted against the pressure. Here, the result of this work is compared to the values of an earlier work of Derstroff [12] from this group. In this study the original approach for gas injection with HNO_3/He mixtures in the stainless steel test gas cylinders was applied. Moreover, literature data of Brown *et al.* [10] is shown in figure 4.11. This study is only one of numerous publications on reaction R1. However, it delivers the most comprehensive data set and their fit of the pressure and temperature

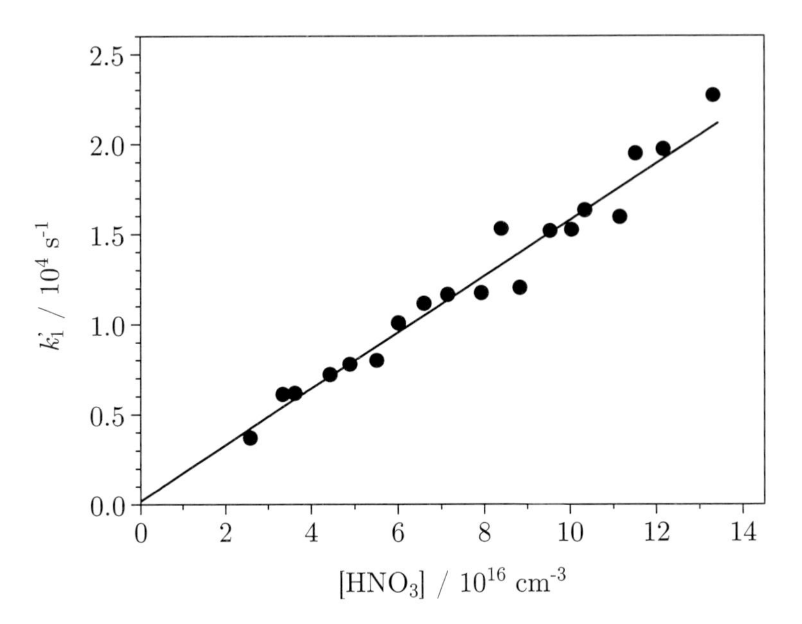

Figure 4.10: Plot of k_1' against $[HNO_3]$.

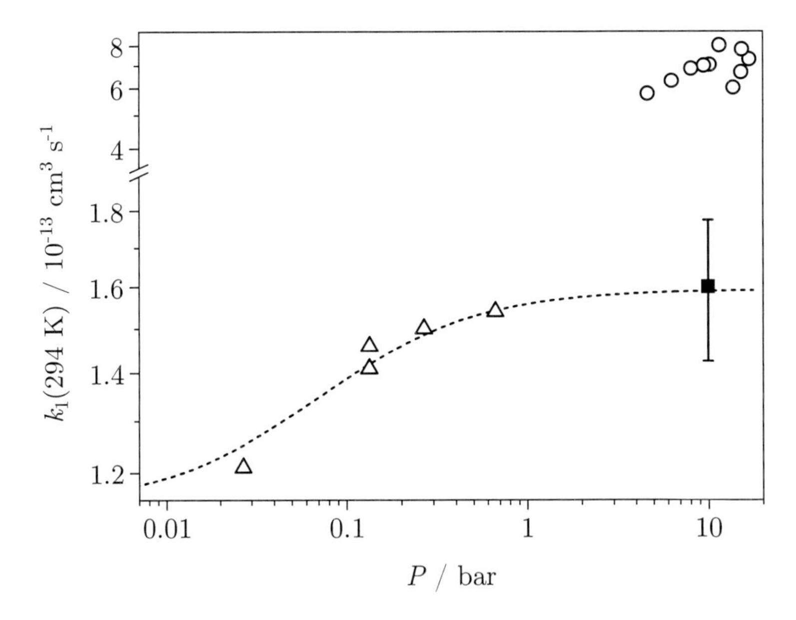

Figure 4.11: Plot of k_1 at room temperature against the pressure: this work (■); Derstroff [12] (○); Brown *et al.* [10]: experiment (△), fit according to a Lindemann-Hinshelwood expression (- -). Note that the ordinate is interrupted between $1.9 \cdot 10^{-13}$ and $3.5 \cdot 10^{-13}$ cm^3 s^{-1}.

dependence of k_1 constitutes the recommendation of ref. [27]. Hence, only this study is consulted here for comparison. Experimental studies on the high-pressure limit of k_1 have not been published yet.

The rate coefficients determined in ref. [12] are by a factor of 4 to 5 higher than the one obtained in this work. This confirms the findings in the UV/Vis absorption studies presented in this chapter regarding the impurities of HNO_3 in gas mixtures. Due to these impurities, not only nitric acid reacts with the OH radicals in the reaction mixture but also the nitrogen dioxide. Thus, the OH radical consumption is significantly accelerated.

The value from this work is in a very good agreement with the predicted high-pressure limit of Brown *et al.* This leads to the conclusion that, with the experimental approach presented here, reliable measurements of the kinetics of reaction R1 at pressure above 1 bar are possible. In contrast, the original experimental approach is not suitable.

4.5.2 Measurements with HNO_3-Containing Gas Mixtures

A general aim of the revision of the gas injection setup was to enable a flexible and independent variation of the concentrations of the different components injected into the reaction cell. With this approach, systematic errors due to reactive impurities of the radical precursor can be avoided, as it is explained in section 4.3.

With the original approach for gas injection such an independent variation was not possible. Nevertheless, the NO_2 impurity in nitric acid gas mixtures must not necessarily cause a non-negligible error in the measured rate coefficients with this approach. Many other factors, like e.g. the age of the gas mixtures, the purity of HNO_3 after synthesis, the rate coefficient of the reaction of interest and the chosen reactant concentration, contribute to the influence of the side reaction NO_2 + OH on the overall measured rate coefficient.

However, the example of the system DEE(-d10) + OH (reaction R3) shows that a significant systematic error can arise. In figure 4.12 the rate coefficients of these reactions measured with the original approach for gas injection in earlier works of this group [13, 14] and the ones obtained with the revised setup in this work are compared. In all measurements gas mixtures in cylinders were used for gas supply.

The rate coefficients at room temperature determined with the original approach for gas injection are between 30 and 35 % higher than the ones obtained in this work. The deviation decreases with increasing temperature. At 550 K it is nearly negligible in the case of k_3, while in the case of k_{3d} a deviation of approx. 15 % remains. [13, 14] These observations can be explained with the significant contribution of NO_2 + OH on the overall measured rate coefficient in the former measurements. The decrease of its influence with increasing temperature might result from the pronounced negative temperature dependence of this side reaction [27]. Because the exact concentration of NO_2 is not known, the consideration of the pseudo-first order rate coefficient of NO_2 + OH in the data evaluation of the former

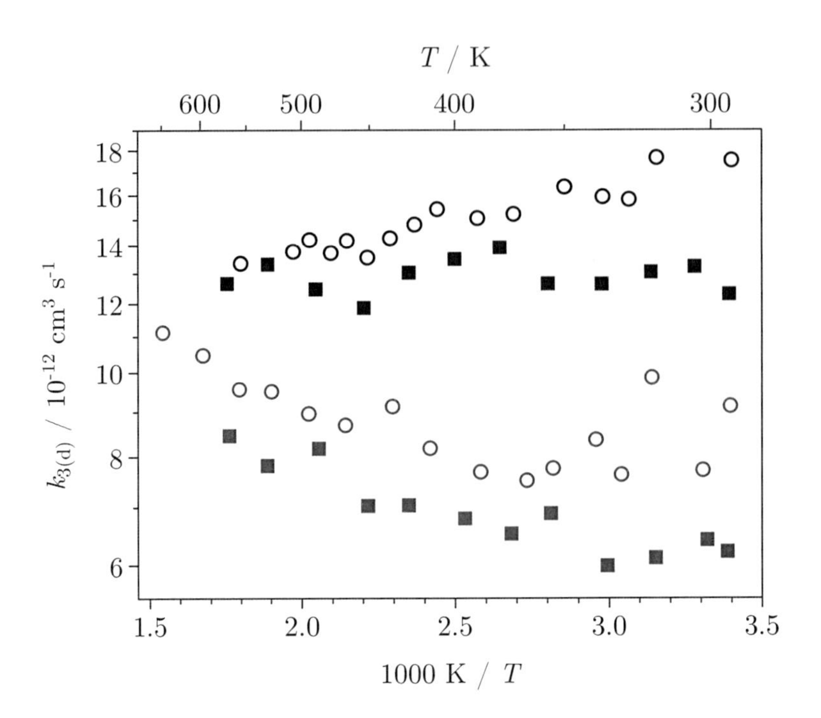

Figure 4.12: Arrhenius plot of k_3 (black) and k_{3d} (red) measured in this group: this work (■); Hetzler [13] (O); prior diploma thesis of the author [14] (O).

measurements is not possible.

The rate coefficients $k_{3(d)}$ obtained with the revised setup for gas injection in this work show a good agreement with literature data, which is discussed in detail in chapter 6. As a result, it can be assumed that with the revised setup for gas injection, the problem of a systematic error caused by the NO_2 impurity in gas mixtures containing HNO_3 can be overcome.

4.5.3 Measurements with Low Reactant Concentrations

Regarding the application of gas mixtures in cylinders for gas supply, another improvement could be achieved with the revision of the gas injection setup, which is independent of the NO_2 impurity of HNO_3. For the study of very fast bimolecular reactions with the PLP/LIF setup applied in this work, low reactant concentrations are needed so that the reaction does not proceed too fast for the possible temporal resolution of the experiment. With the original approach for gas injection the reactant concentration could only be controlled by the filling pressure in the mixture. Consequently, if very fast reactions were studied, very low reactant pressures had to be applied for mixture preparation. With the revision of the gas injection setup the gas mixtures from the cylinders can be further diluted inside the apparatus. Thus,

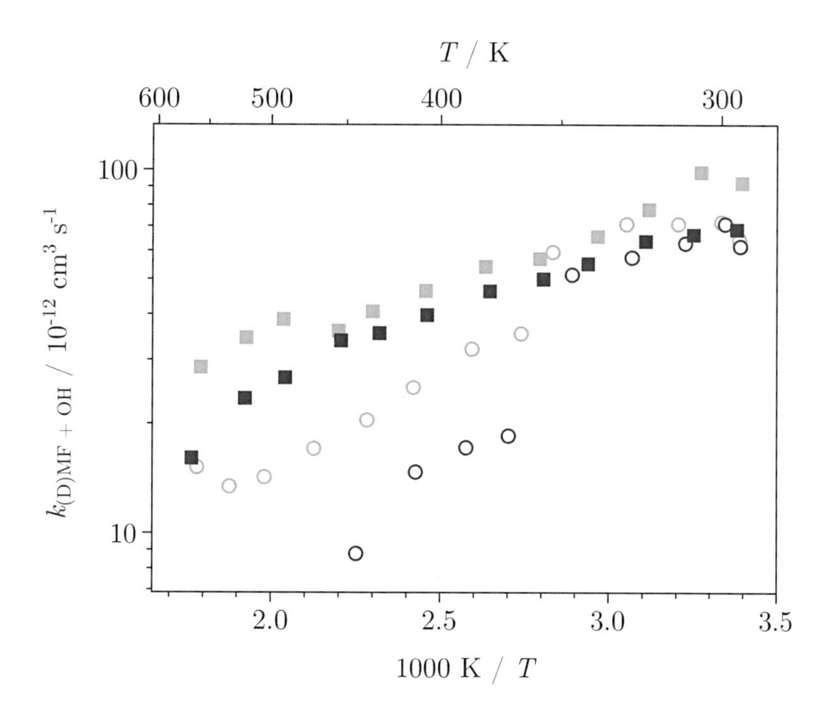

Figure 4.13: Arrhenius plot of $k_{\mathrm{MF\,+\,OH}}$ (blue) and $k_{\mathrm{DMF\,+\,OH}}$ (green): Eble and Owen, recent measurements [120, 135] (■); Eble, diploma thesis [134] (○).

even in the case of the study of fast reactions, moderate reactant pressures can be filled in when a gas mixture is prepared.

In the case of the studies of the systems 2-methylfuran (MF) + OH and 2,5-dimethylfuran (DMF) + OH very low reactant concentrations were needed. They were carried out with the original approach for gas injection in a diploma thesis by Eble [134] and very recently with the revised setup [120, 135] by Eble and Owen. The resulting rate coefficients are compared in figure 4.13.

In the case of the measurements in which the original setup for gas injection was applied significant steps in the Arrhenius behavior between the values at approx. 450 and 470 K were observed. These results were first interpreted in terms of the different reaction channels, which are typically involved in reactions of aromatic compounds with OH radicals [134]. However, when the measurements were repeated with the revised setup for gas injection the results at lower temperatures could be reproduced well, while the steps at higher temperatures disappeared. It turned out that, in the former measurements, new gas mixtures were applied in the measurements above 460 K. Hence, it can be assumed that the low filling pressures applied in the former experiments are connected with a non-negligible systematic error. This error changes from mixture to mixture and, thus, can cause such steps in the measured rate coefficients. Consequently, it can be concluded that filling pressures below approx. 0.1 mbar do not exhibit a sufficient accuracy for the PLP/LIF measurements if the mixtures are prepared with the setup for gas mixture preparation applied in this work. The filling pressures of the reactants in the gas mixtures, which were used in the recent measurements, were significantly higher than in the former measurements due to the further dilution in the PLP/LIF setup. They ranged from approx. 1 to 4.5 mbar. Under these conditions, the error of the pressure is assumed to be small.

As a result, it can be concluded that fast bimolecular reactions can be measured more precisely with the revised setup for gas injection compared to the original one. Even lower concentrations than in the measurements of (D)MF + OH can be achieved when the overall bath gas pressure of a mixture is further increased, while the same filling pressures of the reactant are applied.

4.6 Outlook

The investigation of the impurities of nitric acid reveals the wide range of possibilities which the UV/Vis experiment offers for similar purposes. The setup constructed in this work, generally enables measurements under high-pressure and slow-flow conditions.

For the identification of compounds and studies in a wide spectral range rough measurements can be conducted with a low averaging time, which are fast and easy to realize. For fine measurements higher demands on the baseline stability have to be met. With a self-constructed mount for the absorption cell and the test of different adjustments of the spectrometer, the baseline stability was optimized. Nevertheless, a slight linear baseline drift was observed over an observation period of one hour. With the fitting procedure described in subsection 4.4.3, this problem could be overcome and very sensitive measurements of $[NO_2]$ could be carried out. In the future, this approach of fine measurements for the determination of low concentrations can also be used for other compounds which exhibit a significant absorption band at appropriate wavelengths.

In the application of the high-pressure bubbler, its general practicability and benefits could be demonstrated. The use of this apparatus with arbitrary components is possible in the future, which offers several advantages. On the one hand, the relatively elaborate procedure of mixture preparation can be avoided. On the other hand, it enables the evaporation of compounds into the gas phase which are generally not suitable to be filled into gas cylinders. This can result from a low stability of the compound, as it is the case for nitric acid, as well as from a low vapor pressure. Hydrogen peroxide is another substance which is predestined for the application in a bubbler. The advantages of this OH precursor are discussed in section 4.1. A future test of its suitability for the PLP/LIF approach used in this work could possibly reveal a good alternative for HNO_3 as OH precursor for high-pressure measurements.

For the specific application of the high-pressure bubbler for the supply of pure nitric acid in the gas phase and the measurements on the kinetics of reaction R1, the suitability could be proven in this work. The general experimental approach for the determination of k_1 with PLP/LIF under high-pressure conditions was demonstrated in a first test measurement. For measurements at lower temperatures than room temperature, the concentration of HNO_3 has to be reduced compared to these measurements to avoid condensation of HNO_3 inside the reaction cell. Here, the use of the additional helium line for a further dilution as well as the cooling of the liquid inside the bubbler might become necessary. For the detection of the low HNO_3 concentration an exchange of the lamp in the absorption unit will be required. Here, the application of a low-pressure, cold-cathode mercury lamp at 185 nm is recommended. For this purpose the monochromator has to be exchanged, too. A vacuum monochromator for wavelengths below 200 nm would be suitable. All these necessary changes for measurements with low HNO_3 concentration were already implemented in the high-pressure PLP/LIF setup during this work.

The application of gas mixtures, prepared in test gas cylinders, for the PLP/LIF measurements was examined. The revision of the gas injection setup enables an independent variation of the concentrations of the different compounds and the overall pressure, also when gas mixtures in cylinders are used. Consequently, irregularities in the measurements in general can be investigated more easily. With the determination of the bimolecular rate coefficients from plots of the pseudo-first order rate coefficients against the reactant concentration with all other parameters held constant, many systematic error sources can be avoided compared to the original setup for gas injection. A remaining issue for a future examination is the additional contribution of the errors of the flow controllers to the error of the reactant concentration (cf. subsection 3.2.5). With a general study of systematic (calibration) errors of the controllers, a better assessment of the influence of this error source could be achieved and a procedure for a possibly necessary regular recalibration could be developed.

In addition, it could be shown that significant systematic errors can be avoided with the revised gas injection setup when low reactant concentrations are needed for PLP/LIF measurements with gas mixtures in cylinders for gas supply. This enables studies of the kinetics of very fast reactions with this setup in the future.

5 The Reaction DME + OH

5.1 Introduction

Dimethyl ether (DME), the smallest homolog in the series of aliphatic ethers, is a colorless gas, which has a long history of application in the chemical industry as precursor substance and propellant gas for aerosol sprays [136]. DME is obtained as a by-product in the industrial methanol production. Additionally, it can be produced by methanol dehydration or in a direct route from synthesis gas. Both methods are based on carbonaceous feedstock, which means that a synthesis from biomass as raw material is achievable. [137, 138]

With the high content of oxygen, the absence of a carbon-carbon bond and the low ignition temperature, the combustion of DME proceeds with negligible soot formation and low production of NO_x. [137, 138] All these aspects, combined with the favorable octane and cetane number, turn DME into a promising candidate for an alternative biofuel with broad application, e.g. in diesel engines, gas turbines, private households and as diesel fuel additive [15].

The main disadvantages arise from material-specific properties. For the storage of the gaseous substance in pressure tanks a new infrastructure has to be built up, which is associated with high cost and safety effort. Moreover, the low viscosity and poor lubricity turned out to be problematic for the operation in conventional diesel engines. However, DME as an alternative fuel has been subject of widespread research in the recent years. Possible solutions for many problems were found and first test series in diesel engines in the transportation sector have already been carried out successfully with promising results. [139, 140]

For the investigation of the ignition and combustion characteristics of a fuel, a detailed chemical mechanism with reliable kinetic data is required. The reaction of DME with hydroxyl radicals (OH)

$$\mathrm{H_3COCH_3 + OH \xrightarrow{k_2} products} \tag{R2}$$

is a key reaction step in the combustion process of DME and the most important degradation reaction of emitted DME in the atmosphere. Thus, for both combustion and atmospheric models, the kinetics of this reaction is of fundamental interest.

Many of experimental studies regarding the rate coefficient of reaction R2, k_2, are available in the literature. Direct time-resolved measurements were carried out using resonance absorption [141, 142], resonance fluorescence [143, 144] and laser-induced fluores-

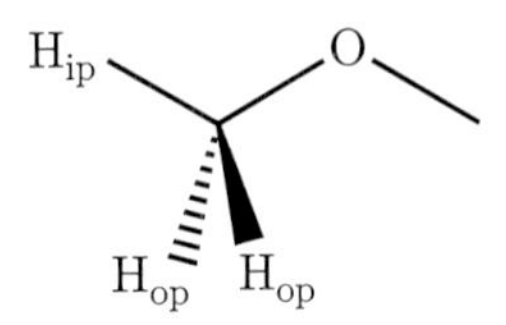

Figure 5.1: Structure of dimethyl ether (DME) with labeling of the different types of H atoms.

cence [95, 96, 100, 101, 118, 145–147]. Furthermore, several relative rate studies are published [142, 148, 149].

The subject of the experimental work of Shannon *et al.* [100, 101] was a low-temperature and pressure regime (63-133 K, 0.3-2.6 mbar), relevant for astrochemical processes. They found a strong positive temperature dependence of the rate coefficient k_2 and also a pronounced positive pressure dependence under these conditions.

Cook *et al.* [141] reported on shock-tube measurements at temperatures from 923 to 1470 K, at which combustion processes take place. The pressure ranged from 1.1 to 3.0 bar. All other experiments in the literature [95, 96, 118, 142–149] were carried out at moderate temperatures between 195 and 850 K. This is the relevant region for the ignition and atmospheric chemistry, which was also the objective of this work. The pressures ranged from 0.0026 to 1 bar. The only study dealing briefly with a possible pressure dependence of k_2 at moderate temperatures is the one of Carr *et al.* [96]. They carried out measurements between 2.6 and 330 mbar and found k_2 to be independent of the pressure under these conditions.

Except for the values of Perry *et al.* [143], which are slightly higher than the others, there is a good agreement between the numerous rate coefficients available in the literature. Atkinson assumed in his review [85] that k_2 is overestimated in the work of Perry *et al.* due to reactive impurities in the reaction system. The Arrhenius plot of k_2 shows a positive temperature dependence with a pronounced curvature in the temperature range between 195 and 1470 K.

There are also several theoretical studies dealing with the kinetics of reaction R2 [96, 101, 150–156]. For the calculation of the structures, vibrational modes and energies of the reactants and transition states, several different quantum chemical methods were used. All of them showed that the H-atoms can be abstracted at the in-plane (ip) and out-of-plane (op) position. This differentiation is illustrated in figure 5.1.

The recent studies predict the formation of a hydrogen-bonded pre-reaction complex in the main channel of the reaction, which can either decompose or react in an intra-molecular hydrogen abstraction [96, 151–153, 155, 156]. The so-formed product complex decomposes fast into the reaction products:

$$H_3COCH_3 + OH \rightleftharpoons C_2H_6O\cdots HO \quad \text{(R11)}$$

$$C_2H_6O\cdots HO \longrightarrow C_2H_5O\cdots HOH \quad \text{(R12)}$$

$$C_2H_5O\cdots HOH \rightleftharpoons H_2COCH_3 + H_2O. \quad \text{(R13)}$$

Beyond that, several discrepancies in the results of the different theoretical studies exist. Especially the use of density functional theory (DFT) methods is discussed controversially. Bottoni *et al.* pointed out "the inadequacy of the DFT approach, based on the most popular functionals available in literature (B3LYP, B3PW91, BLYP), to describe this kind of reaction characterized by polar transition states" [150]. The reason for this conclusion was the fact that no transition state (TS) could be found at B3PW91/6-31G* and BLYP/6-31G* level of theory and only TS_{op} (TS of the abstraction of H_{op}) at the B3LYP/6-31G* level. El-Nahas *et al.* [153] did not find any transition states with the B3LYP functional. Moreover, the rate coefficient based on the calculations at BH&HLYP/6-311++G(d,p) level of theory disagreed with the experimental findings. As a result they concluded: "Therefore, BH&HLYP/6-311++G(d,p) optimized geometries are not recommended for the considered reactions or its results have to be taken with caution" [153].

Nevertheless, Ogura *et al.* [154] used the B3LYP functional for the validation of their group rate expressions for reactions of OH radicals with ethers. Zavala-Oseguera *et al.* [155] carried out a theoretical study of reaction R2 at different levels of theory including coupled cluster and DFT methods. They found the best agreement with the experiment for the rate coefficients determined on the basis of the quantum chemical calculations at M05-2X/6-311++G(d,p) level of theory. As a result, the M05-2X functional was chosen for the description of the reactions of higher ether homologs with OH. However, the mechanistic findings for reaction R2 with M05-2X/6-311++G(d,p) differ from the other publications. The abstraction of H_{ip} was obtained as the main channel, whereas most of the other theoretical studies [96, 150–153, 156] predict the abstraction of H_{op} energetically favored. Zhou *et al.* [156] computed the geometries of reaction R2 at BH&HLYP/6-31G(d) level of theory and did not report about any profound problems in the application of this DFT functional either. Moreover, in good agreement with their computation at MP2/6-311G(d,p) level of theory and with other theoretical studies [96, 150–153], the calculations yielded the energy of TS_{op} lower than the one of TS_{ip}.

For the higher homologs of ethers the applicability of more reliable, but computationally demanding, methods like coupled cluster theory is limited. Thus, a detailed examination of the potential of common DFT functionals for the description of reactions of ethers with OH radicals would be helpful.

The rate coefficient of reaction R2, k_2, was studied with different theoretical approaches

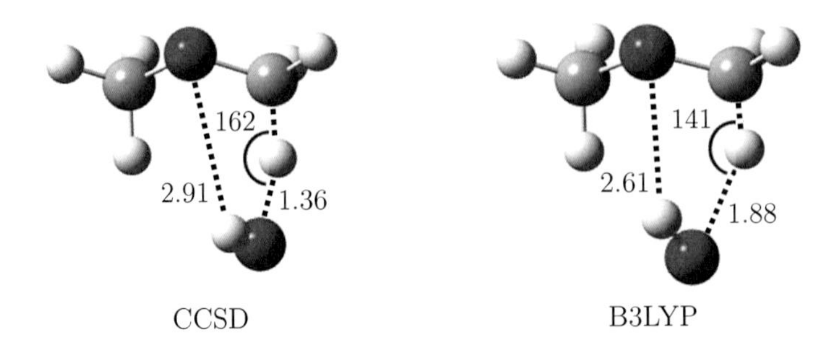

Figure 5.2: Transition states calculated at CCSD/cc-pVDZ and B3LYP/6-311G(2d,d,p) level of theory (bond lengths in Å; angles in degrees). The picture is partly taken from ref. [157].

[96, 101, 151–156]. A good agreement between experiment and theory was obtained in the recent studies of Zhou *et al.* and Carr *et al.*, who considered the unimolecular reaction steps of the prereactive complex in the op-channel explicitly. However, a potential pressure dependence and pre-equilibrium conditions were not examined.

In a cooperation within our group an extensive study of reaction R2 was carried out. The results are already published in ref. [157]. The goal was to resolve the discrepancies in the calculations of k_2 depending on the quantum chemical method and to investigate the reaction mechanism in detail to gain more information about the validity of the pre-equilibrium assumption.

Quantum chemical calculations were carried out by Kiecherer and Szöri [18,157], amongst others, with CCSD(T)/cc-pV(T+Q)Z//CCSD/cc-pVDZ and CBS-QB3. The latter is a compound method, in which the geometries and frequencies are calculated at B3LYP/6-311G(2d,d,p) level of theory. Consistently, an indirect channel proceeding via a prereactive complex was obtained for the abstraction of H_{op}. TS_{ip} could only be located in the calculations at CCSD/cc-pVDZ level of theory indicating a complex-forming mechanism for this channel as well, with a higher barrier for the abstraction compared to the op-channel. At B3LYP/6-311G(2d,d,p) level of theory no transition state was found for the ip-channel.

Fundamental differences were observed in the structures for TS_{op} obtained from the different quantum chemical methods. A comparison is shown in figure 5.2. In the geometry calculated at CCSD/cc-pVDZ level of theory the hydrogen bond between the hydroxyl radical and the ether is stretched more, the forming OH bond is shorter, and the OHC angle is considerably bigger compared to the one calculated with B3LYP/6-311G(2d,d,p). Thus, the reaction coordinate in the first case is very close to a translation of the reacting hydrogen atom towards the oxygen atom of the hydroxyl radical, while in the second one it corresponds to a rotational motion of the oxygen atom towards the hydrogen atom. As a result, the imaginary wavenumbers differ significantly from each other. The calculations at

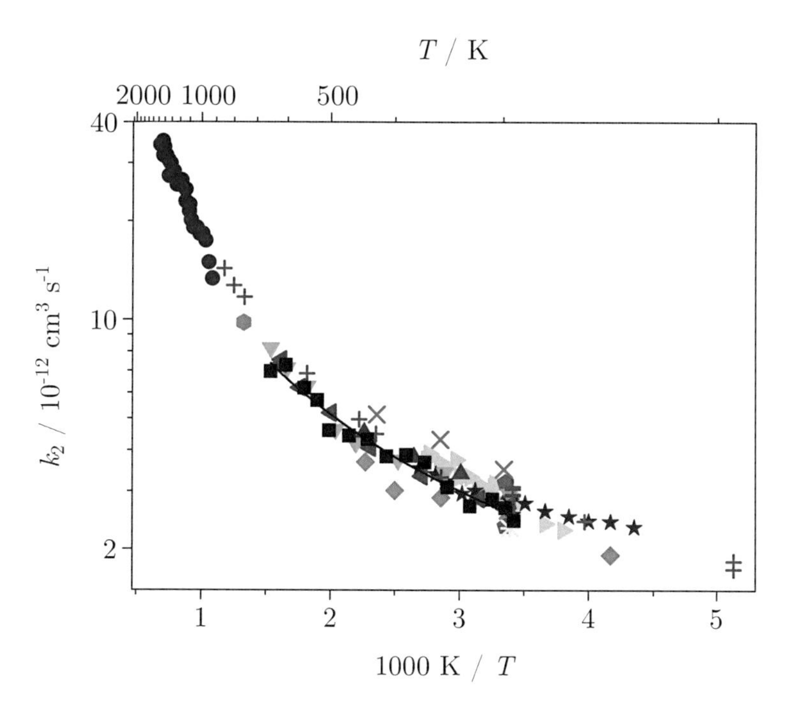

Figure 5.3: Arrhenius plot of experimentally determined k_2: Results of the prior diploma thesis of the author (■) with best fit of a modified Arrhenius expression (—) [14,157], Perry *et al.* [143] (×), Tully and Droege [147] (▲), Wallington *et al.* [144] (◆), Wallington *et al.* [149] (✳), Nelson *et al.* [142] (⬟), Mellouki *et al.* [118] (★), Arif *et al.* [95] (▼), DeMore and Bayes [146] (▶), Tranter and Walker [148] (⬢), Bonard *et al.* [145] (◀), Cook *et al.* [141] (●), Carr *et al.* [96] (+).

CCSD/cc-pVDZ level of theory gave $1354i$ cm^{-1}, in contrast to a value of $158i$ cm^{-1} at B3LYP/6-311G(2d,d,p) level of theory.

The kinetics of reaction R2 was described on the basis of these results by Kiecherer [18,157] using conventional transition state theory (TST) under the assumption of a preequlibrium for the complex formation. Despite the above-named discrepancies in the predicted mechanistic properties, the obtained k_2 based on the CCSD(T)/cc-pV(T+Q)Z//CCSD/cc-pVDZ and CBS-QB3 calculations both show a satisfying agreement with the experiment. Hence, the adequacy of the assumption of a pre-equilibrium could be validated. However, due to the reasonable agreement of both theoretical results with the experiment, it could not be found out which calculations give an inadequate description of TS_{op}.

The fundamental differences in the geometries of TS_{op} though must lead to different isotope effects of the zero-point corrections and the tunneling contributions. As a result, the isotope effect of k_2 was expected to deviate significantly depending on the quantum chemical method. Hence, an analogous study was conducted for the rate coefficient k_{2d} of the reaction DME-d6 + OH by Kiecherer and Szöri [18,157].

Indeed, the calculated isotope effect of k_2 characterized by the ratio k_2/k_{2d} ranged from 4.1 at 400 K to 2.9 at 550 K at CCSD(T)/cc-pV(T+Q)Z//CCSD/cc-pVDZ level of theory, while values between 1.3 and 1.2 were obtained at the CBS-QB3 level.

Although a high reliability can be assumed for the calculations at CCSD(T)/cc-pV (T+Q)Z//CCSD/cc-pVDZ level of theory, a further validation of the theoretical findings was expected from an experimental investigation of the isotope effect of k_2. No experimental values for k_{2d} were available in the literature at that time.

In the prior diploma thesis of the author [14] k_2 was measured at pressures from 5.9 to 20.9 bar. No pressure dependence was found. Moreover, the measurements were carried out in a temperature range between 292 and 650 K. A comparison of the obtained rate coefficients at different temperatures with the values from the literature is shown in figure 5.3. The remarkable agreement proves the reliability of the experimental approach with respect to the investigation of the system DME(-d6) + OH.

In the present work, the Arrhenius behavior of k_{2d} was measured following the experimental procedure already tested in ref. [14]. With the results, the adequate transition state structure obtained with the particular quantum chemical method was found and a further validation of the theoretical results could be achieved.

Meanwhile, Carr *et al.* [96] published a study on the isotope effect of the rate coefficient of reaction R2. Their results for k_{2d} are compared to the findings in this work in section 5.4.

5.2 Experimental Procedure

The measurements of k_{2d} were realized with the original approach for gas injection, which is described in subsection 3.1.1 and further discussed in chapter 4. Experiments were conducted at pressures between 13.0 and 20.4 bar at six different temperatures in a range from 387 to 554 K, while the maximum temperature difference between gas inlet and outlet of the reaction cell amounted to 9 K (cf. section 3.2). The concentration of DME-d6, [DME-d6], ranged from $1.0 \cdot 10^{17}$ to $1.7 \cdot 10^{17}$ cm^{-3}, while $[HNO_3]$ varied between $7.8 \cdot 10^{15}$ and $1.1 \cdot 10^{16}$ cm^{-3}.

At each temperature, one series of measurements consisting of 20 LIF intensity-time profiles was conducted. In the applied experimental approach the overall pressure and the concentration of all components were dependent on the filling level of the reservoir gas cylinder. Hence, a pressure loss between 2.2 and 3.1 bar during one series of measurements was observed. According to the ratio between the present pressure and the pressure at the beginning, the concentrations of all components changed during one series of measurements as well. However, the pressure and concentration change during one single measurement never exceeded 1.5 % and, thus, are assumed to be negligible in consideration of the overall experimental error (cf. subsection 3.2.5).

In an extra series of measurements of the system DME + OH in the prior diploma thesis of the author [14], a potential pressure dependence for k_2 could be excluded and the linear relation between the pseudo-first order rate coefficient k_2' and [DME] could be proven. Here, at one fixed temperature 61 measurements were carried out in a broad pressure and [DME] range. Due to the restricted availability of DME-d6, a realization of a corresponding analysis for k_{2d} was not possible. However, due to the similarity of the system itself, and of the measuring conditions, the results for DME + OH should be transferable on the system DME-d6 + OH.

The gas flow was set to 2 slm for all measurements. A potential gas flow dependence of the observed reaction rate was tested roughly at room temperature preliminary to the actual measurements. Here, several LIF intensity-time profiles were recorded at different gas flows between 0.2 and 10 slm and a potential dependence of k_{2d}' on this parameter was examined. No correlation was observed.

The laser fluence of the photolysis laser amounted to maximum 4 mJ cm^{-2}, while the one of the excitation laser ranged from approx. 15 to 30 mJ cm^{-2}. The problem of a slight decalibration of the dye laser is discussed in section 3.2. In the case of the study of k_{2d} the excitation wavelength was set to 281.923 nm. For the detection of the fluorescence radiation the monochromator wavelength was set to 308 nm and the slit to 2 mm. This corresponds to a bandwidth of 16 nm (fwhm).

The repetition rate of the experiment amounted to 5 Hz. Approx. 80 data points at different time delays between photolysis and detection were recorded for the characterization of one LIF intensity-time profile. Each of these data points consisted of ten single values at a given time delay averaged internally by the measuring program (cf. section 3.2). Care was taken that the measured signal reached the zero baseline, before the recording was stopped.

Three different gas mixtures were applied in the measurements of k_{2d}, containing DME-d6, HNO_3, and helium. The age of the mixtures ranged from 1 to 8 days. Prior to the experiments, HNO_3 was freshly prepared according to the procedure described in section 3.1 and stored in darkness. Thus, the age of the HNO_3 applied in the gas mixtures did not exceed 21 days. The following substances were used: DME-d6 99 %, Sigma Aldrich; H_2SO_4 98 %, Roth; $KNO_3 \geq 99$ %, Roth; He > 99.999 %, Air Liquide.

5.3 Analysis

With the deeper insight gained from the studies of the purity of HNO_3 (cf. chapter 4), the analysis of the measurements regarding the system DME-d6 + OH of ref. [157] was revised. No potential side reactions were taken into account.

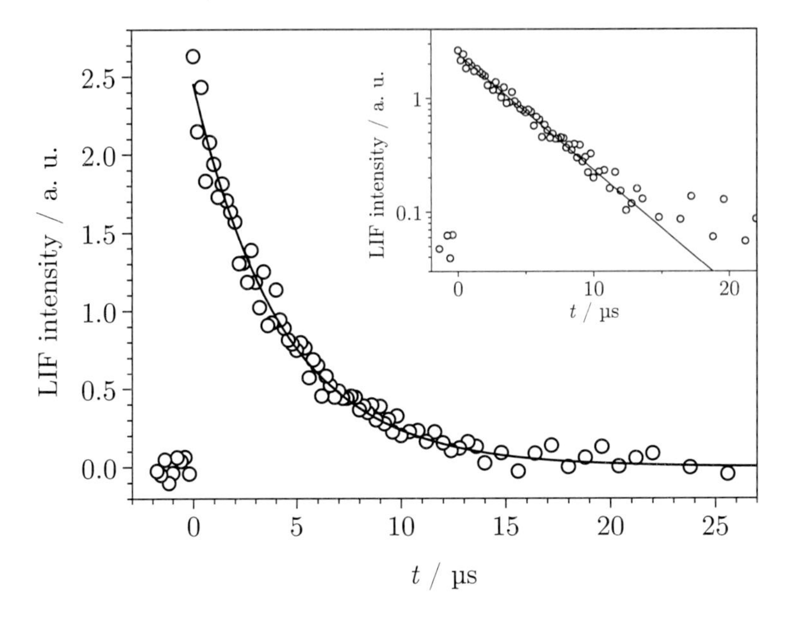

Figure 5.4: LIF intensity-time profile of the reaction DME-d6 + OH at 17.3 bar, 554 K, and [DME] = $1.25 \cdot 10^{17}$ cm^{-3}. k'_{2d} was determined from the best fit to the linearized signal, which is shown as well.

A typical LIF intensity-time profile of the system DME-d6 + OH is depicted in Figure 5.4. Without exception, the measured profiles could be described adequately with a mono-exponential decay function. The zero baseline was determined by averaging the data points recorded at a negative time delay between photolysis and detection. Then, the pseudo-first order rate coefficients k'_{2d} were obtained from the slope of the best fit to the linearized signals according to

$$\ln(I) = \ln(I_0) - k'_{2d} t \,. \tag{5.1}$$

An overview of all obtained k'_{2d} and the corresponding measuring conditions is given in appendix A.2. The bimolecular rate coefficients k_{2d} were calculated with the following expression:

$$k_{2d} = \frac{k'_{2d}}{[\text{DME-d6}]} \,. \tag{5.2}$$

By assuming ideal gases, the concentration of the reactant [DME-d6] was determined from its initial partial pressure $P_{\text{DME-d6,mix}}$, the overall pressure P_{mix} in the particular mixture, and the current total pressure in the reaction cell P according to

$$[\text{R}] = \frac{P_{\text{DME-d6,mix}}}{P_{\text{mix}}} \frac{P}{\text{R}T} \text{N}_\text{A} \,. \tag{5.3}$$

The assumption of pseudo-first order conditions could be validated by estimating the initial OH radical concentration $[OH]_0$ with equations 3.2 and 3.1. Values between $0.9{\cdot}10^{12}$ and $1.6{\cdot}10^{12}$ cm^{-3} were obtained. Thus, the ratio $[OH]_0$/[DME-d6] never exceeded 0.001 %.

As already discussed in subsection 3.2.5 the influence of several sources of error can be tested by the examination of potential correlations of the experimental results with different parameters. First, a gas flow dependence of k'_{2d} could be excluded with the procedure described in section 5.2. Second, the difference between the determined bimolecular rate coefficients and the values obtained from the best fit of the modified Arrhenius expression (cf. section 5.4) was plotted against the age of the applied gas mixture. No systematic dependence was observed. Third, a correlation of this difference with the laser fluence of the detection laser was investigated and could be excluded. A potential dependence on the fluence of the photolysis laser in order to examine the influence of OH recombination was not explicitly tested. However, the investigation of the system DME-d6 + OH was conducted at low OH concentrations and high pseudo-first order rate coefficients compared to the systems DEE(-d10) + OH and DMM + OH (cf. chapter 6 and 7). As no influence of the laser fluence and thus of OH recombination was observed there, this error source can also be excluded in this case through conclusion by analogy.

5.4 Results and Discussion

20 bimolecular rate coefficients were determined at each temperature and different pressures. All k_{2d} were plotted against the pressure. No correlation in the narrow pressure ranges (cf. section 5.2) resulted. Moreover, a potential pressure dependence of k_2 in a broad pressure range was excluded in the prior diploma thesis of the author [14]. This can also be seen as further evidence for the independence of k_{2d} from the pressure under the examined conditions. Consequently, the mean of all bimolecular rate coefficients at a particular temperature was taken for the further evaluation. The obtained values are listed in table 5.1.

Figure 5.5 shows the Arrhenius plot of $k_{2(d)}$ determined in the prior diploma thesis of the author and in this work, respectively. Moreover, a comparison to the experimental values of

T / K	$\overline{k_{2d}}$ / 10^{-12} cm^3 s^{-1}
388 ± 1	1.0 ± 0.3
417 ± 1	1.0 ± 0.3
434 ± 1	1.1 ± 0.3
498 ± 1	1.5 ± 0.5
527 ± 1	1.6 ± 0.5
554 ± 1	1.8 ± 0.6

Table 5.1: Overview of all averaged rate coefficients $\overline{k_{2d}}$ obtained in this work.

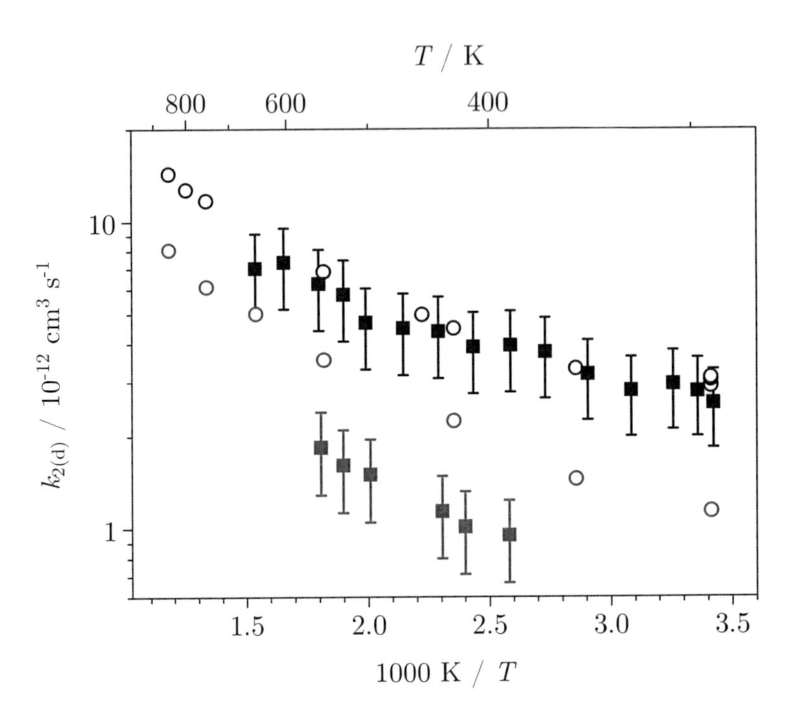

Figure 5.5: Arrhenius plot of experimentally determined k_2 (black) and k_{2d} (red): prior diploma thesis of the author [14] (■); this work (■); Carr *et al.* [96] (○). The error bars represent the estimated error of 30 %.

Carr *et al.* [96] is given.

The remarkably good agreement between the obtained k_2 has already been discussed in section 5.1. However, a deviation in the values for k_{2d} can be observed. The rate coefficients obtained by Carr *et al.* at a pressure of 67 mbar are up to a factor of two higher than the ones measured in this work at elevated pressures. A reason for this discrepancy could not be found in this work. Due to the analogy to the reaction of the undeuterated species, an actual negative pressure dependence can be excluded. The most obvious error source in the experimental approach applied in this work for the study of k_{2d} is the neglect of any side reactions, as discussed in subsection 3.2.5 and chapter 4. In the case of an influence of side reactions though, the measured values always overestimate the actual rate coefficient. The fact that the values determined in this work are lower than the ones of Carr *et al.* argues against an error caused by side reactions in this work.

In conformance with the system DME + OH, k_{2d} was parametrized with the best fit of a modified Arrhenius expression given by

$$k_{2d} = 7.27 \cdot 10^{-23} \left(\frac{T}{\mathrm{K}}\right)^{3.568} \exp\left(\frac{780\ \mathrm{K}}{T}\right)\ \mathrm{cm^3\ s^{-1}}\,. \tag{5.4}$$

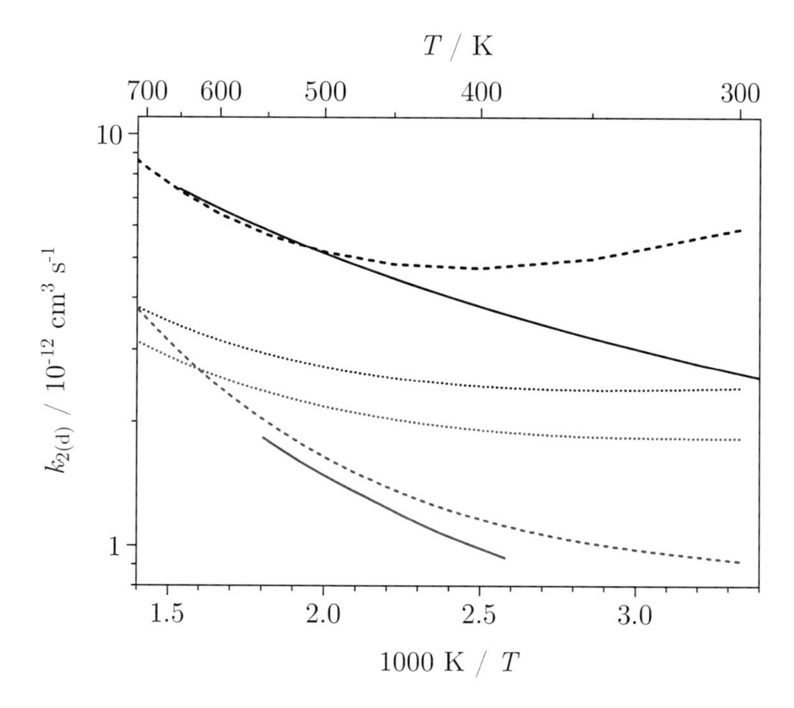

Figure 5.6: Arrhenius plot of k_2 (black) and k_{2d} (red): Arrhenius fit to experimental results from the prior diploma thesis of the author [14] (—) and from this work according to equation 5.4 (—); calculated high-pressure limiting rate coefficients at CCSD(T)/cc-pV(T+Q)Z//CCSD/cc-pVDZ (- -) and CBS-QB3 (····) level of theory from refs. [18, 157].

In figure 5.6 a comparison between the experimentally determined Arrhenius behavior of $k_{2(d)}$ and the theoretical results of Kiecherer [18, 157] is depicted.

With both methods slight deviations of the theoretical results for k_2 from the experimental findings can be observed. At CCSD(T)/cc-pV(T+Q)Z//CCSD/cc-pVDZ level of theory a good agreement at temperatures above 450 K is obtained, while the experiment is overestimated at lower temperatures. Here, a transition to a negative temperature dependence is predicted by the theory, whereas the experimental values continue to increase with the temperature in this region. A positive temperature dependence in the whole temperature range is predicted at the CBS-QB3 level. However, the slope of the Arrhenius behavior is underestimated, resulting in a divergence of the experimentally and theoretically determined rate coefficients with increasing temperature. Nevertheless, taking into account that the quantum chemical results were applied without any adjustment to the experiment, the agreement is reasonable for both methods.

However, as already discussed in section 5.1, a considerable discrepancy in the transition state structures and reaction coordinates in the main channel was observed for the different

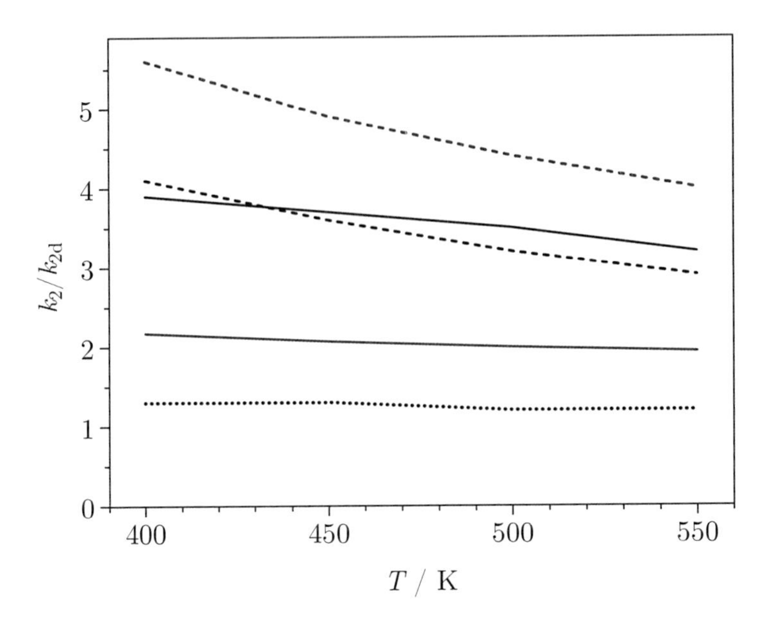

Figure 5.7: Temperature dependence of the isotope effect k_2/k_{2d}. Results from this group (black): experiment, this work (—), calculations at CCSD(T)/cc-pV(T+Q)Z// CCSD/cc-pVDZ level of theory [18, 157] (- -), calculations at CBS-QB3 level of theory [18, 157] (····); Results of Carr *et al.* [96]: experiment (—), theory (- -).

theoretical levels. A validation of the predicted mechanistic properties can only be achieved by the examination of the obtained isotope effect. Consequently, this quantity is examined in detail in the following.

The experimental and theoretical results for the isotope effect, characterized by the ratio k_2/k_{2d}, of this group and of Carr *et al.* are depicted in figure 5.7. For the determination of the isotope effect of the experimental values of Carr *et al.*, the best fit of the modified Arrhenius expression to their results for k_{2d} was calculated. A parametrization of k_2 was reported in ref. [96]. With these expressions the ratio k_2/k_{2d} could be determined in the considered temperature range. The theoretically obtained isotope effect of Carr *et al.* is directly given in ref. [96].

Due to the above-named discrepancy between the experimental values for k_{2d}, the experimentally determined isotope effects of Carr *et al.* and this work also differ considerably. Significantly lower values are obtained from the experiments of Carr *et al.* However, their theoretical results, also carried out with high level quantum chemical methods, indicate a somewhat higher ratio k_2/k_{2d} than the one obtained in this work, validating the experimental results for k_{2d} herein.

By considering the theoretical findings obtained within this group, a very good agreement of the results at CCSD(T)/cc-pV(T+Q)Z//CCSD/cc-pVDZ level of theory with the experiment can be observed with a deviation of 10 % at most. In contrast, with a discrepancy of at least 60 %, the values on the basis of the calculations at the CBS-QB3 level disagree significantly. Consequently, the inadequacy of the theoretical description of the reaction coordinate and transition state structures at B3LYP/6-311G(2d,d,p) level of theory as described in section 5.1 can be stated, while the calculations at the CCSD(T)/cc-pV(T+Q)Z//CCSD/cc-pVDZ level exhibit a high reliability.

5.5 Conclusion and Outlook

Beside the study of Carr *et al.* [96], which was published at the same time, the rate coefficient k_{2d} of the reaction DME-d6 + OH was first characterized in the present work. The results disagree with the values reported in ref. [96] by a factor of up to two. No explanation for this discrepancy could be found. However, the theoretical investigation of k_{2d} based on high level quantum chemical calculations carried out within this group [18, 157] and by Carr *et al.* [96] indicate the adequacy of the experimental results obtained in this work.

Generally, due to the large amount of experimental studies on k_2 in a broad temperature range and their generally very good agreement, the system DME + OH is predestined for kinetic benchmark studies. In this work and in ref. [18] (published together in ref. [157]), it could be shown that the investigation of the kinetic isotope effect can deliver important additional information about the examined system. Hence, for future studies of this type further validation of the experimental results for k_{2d} would still be desirable.

To evaluate the performance of the B3LYP functional in the description of the system DME + OH, the isotope effect of k_2 was examined in detail. The calculations of k_2/k_{2d} on the basis of the results at CBS-QB3 level of theory [18, 157] yielded values which underestimate the experimental findings significantly. This leads to the conclusion that the calculations at the CBS-QB3 level, especially with the use of B3LYP/6-311G(2d,d,p) for the geometries, cannot characterize the reaction process of DME + OH satisfactorily. For an adequate description of this type of complex-forming reactions the application of high level *ab initio* methods like CCSD(T)/cc-pV(T+Q)Z//CCSD/cc-pVDZ is recommended.

6 The Reaction DEE + OH

6.1 Introduction

Diethyl ether (DEE) is the second homolog in the series of aliphatic symmetrical ethers after DME. It is a well-known solvent with application in laboratory and industrial processes. Because DEE is formed as a by-product in the synthesis of other chemicals, a direct industrial production of considerable amounts of DEE is not necessary at present. However, due to simple synthesis routes on the basis of (bio-)ethanol, large amounts of DEE could be provided cost-efficiently from renewable resources. [158]

DEE has become famous for the risk of violent explosions in the handling of the substance. On the one hand, this results from its tendency to form explosive peroxides under the impact of light and elemental oxygen [159–161]. On the other hand, with 443 K [158], DEE exhibits an autoignition temperature even lower than the one of DME combined with a low boiling point of 308 K [158]. Consequently, explosive DEE/air-mixtures are formed very fast and hot surfaces possibly suffice for an ignition [17]. This resulted in an application of DEE as a model substance for the investigation of safety-relevant ignition processes on hot surfaces in the field of risk prevention in the industry [162].

In contrast, low ignition temperatures of combustible material are an advantage for the application in diesel engines. The potential of DME as a diesel biofuel has already been discussed in section 5.1. However, some material specific characteristics of DME are disadvantageous [137,138]. Hence, the focus of current biofuel research not only lies in overcoming the problems of the most popular candidates but also in the search for optimized alternative fuels for a specific application.

From this point of view DEE has attracted increased attention in the recent years, especially as a potential additive for blending diesel and biodiesel. It exhibits excellent solving properties in all common organic solvents [158] as well as in diesel fuels [163]. However, although the suitability of DEE as a cold-start aid for diesel engines is long-known [163], the detailed ignition and combustion characteristics are the subject of current research. Recently, numerous studies on the emission and performance characteristics for both DEE-diesel [164–168] and DEE-biodiesel [169–172] blends indicated significantly improved properties due to the addition of DEE.

As a result, a detailed understanding of the ignition behavior of DEE is desirable in both counts, the safety-relevant aspect and the application as a diesel fuel additive. Especially for the low-temperature ignition process the reaction of DEE with OH radicals

$$\mathrm{DEE} + \mathrm{OH} \xrightarrow{k_3} \mathrm{products} \tag{R3}$$

is of fundamental importance. Moreover, this reaction represents the most important degradation channel of DEE in the atmosphere (cf. chapter 1). Consequently, for the modeling of combustion and atmospheric processes the rate coefficient of reaction R3, k_3, should be well characterized under diverse conditions.

However, the kinetic parameters of this reaction are estimated in all recently published oxidation mechanisms of DEE [161, 173–176] because of the poor availability of literature data for k_3 under ignition-relevant conditions, i.e. temperatures above the ignition temperature and pressures higher than 1 bar. Nevertheless, for lower temperatures and pressures numerous experimental studies on the kinetics of reaction R3 have already been published [118, 142, 144, 147, 148, 177–180].

Considering the direct studies [118, 142, 144, 147], a consistent picture emerges. Tully and Droege [147] carried out measurements with pulsed laser photolysis/laser-induced fluorescence (PLP/LIF) in a temperature range between 295 and 442 K at a pressure of 0.53 bar. A slightly negative temperature dependence of k_3 was obtained. These values could be reproduced by Wallington *et al.* [144] in their study with flash photolysis/resonance fluorescence at temperatures between 240 and 440 K and pressures of 0.033 and 0.067 bar. Nelson *et al.* [142] applied pulsed radiolysis/transient UV absorption as a direct technique at 298 K and 1 bar. The results agree well with the above-named studies. PLP/LIF was also used by Mellouki *et al.* [118], who studied k_3 in a temperature range between 230 and 372 K at 0.13 bar. Again, the values of the other direct investigations could be confirmed. Due to the good agreement of all direct studies, it can be derived that k_3 is not dependent on the pressure between 0.033 and 1 bar in the corresponding temperature range.

The results of the indirect experimental studies carried out with different relative rate methods, all at pressures of approximately 1 bar, deviate partly from the findings of the direct measurements. While the single value of Bennett and Kerr [178] at 294 K is in good agreement, their temperature dependent measurements [179] underestimate the values of the above-named publications. The temperature dependence of k_3 was also investigated indirectly by Semadeni *et al.* [180] in a temperature range between 247 and 373 K. The results agree reasonably well with the direct studies below 310 K. At higher temperatures a divergent trend to lower values of k_3 emerges. Lloyd *et al.* [177] and Nelson *et al.* [142] determined single values for k_3 at 305 K and 298 K, respectively, applying relative rate methods. While the first study underestimates k_3 significantly, a good agreement with the direct studies was obtained for the latter.

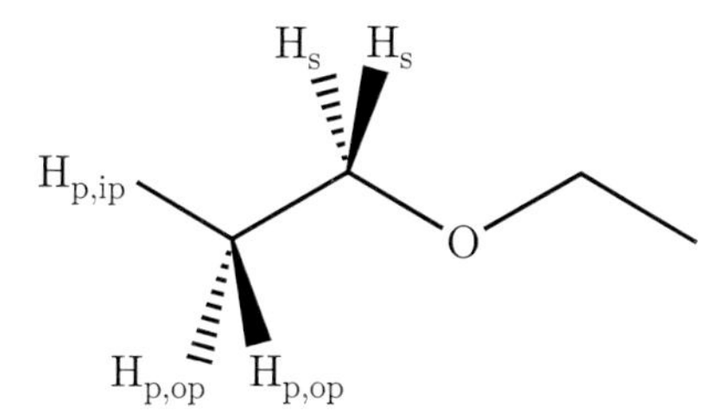

Figure 6.1: Structure of diethyl ether (DEE) with labeling of the different types of H atoms.

The investigation of Tranter and Walker [148] is the only one dealing with temperatures above the ignition temperature of DEE. They determined k_3 at 753 K and 0.67 bar applying a competitive technique as well. A considerably increased rate coefficient in comparison to the direct low-temperature studies was measured. This finding was explained in terms of a strongly curved Arrhenius behavior passing a minimum at approximately 370 K. However, no other studies at high temperatures are published yet to validate these results.

Two theoretical studies have been published [155, 181] regarding the detailed reaction mechanism of reaction R3. Generally, with the different types of H atoms, which are illustrated in figure 6.1, there exist three possible channels for reaction R3. On the one hand, the secondary H atoms H_s in α-position with respect to the ether O atom can be abstracted. On the other hand, an abstraction of the primary H atoms H_p at the β-site can occur. Here, two different conformational types of H atoms can be differentiated: the ones in-plane $H_{p,ip}$ and out-of-plane $H_{p,op}$ with respect to the COC-substructure.

The potential energy surface was characterized by Zavala-Oseguera *et al.* [155] at M05-2X/ 6-311++G(d,p) level of theory considering all three reaction channels in their calculations. Sandhiya *et al.* [181] carried out calculations at UM06-2X/6-31++G(d,p) level of theory. They did not distinguish between the different conformerical types of H_p in β-position. However, in both studies the abstraction of H_s was obtained as the main channel. In line with the findings for DME (cf. section 5.1) it proceeds indirectly via a prereactive and postreactive complex, in which the OH radical is weakly associated with the O atom of the ether through a hydrogen bond. Stabilization energies of the prereactive complex of around 30 kJ mol^{-1} were found.

Zavala-Oseguera *et al.* [155] also investigated the kinetics of reaction R3. They applied interpolated variational transition state theory by mapping, and calculated k_3 at the high-pressure limit between 280 and 2000 K. Although a good agreement between experimental and theoretical values at room temperature was stated, the obtained pronounced positive temperature dependence in the whole explored temperature range is contradictory to the negative temperature dependence observed in the experiments.

However, the existence of multiple and multi-step reaction channels can lead to complicated relations between the rate coefficient and the temperature and pressure. Thus, for a satisfactory understanding of the kinetics of reaction R3 in a broad temperature and pressure range further work is needed.

To investigate k_3 under ignition-relevant conditions a comprehensive theoretical and experimental study was carried out within our group. The theoretical results are partly published in ref. [18].

Quantum chemical calculations were carried out by Kiecherer [18] and Szöri [182] at CCSD(T)/cc-pV(T+Q)Z//CCSD/cc-pVDZ level of theory. All three possible reaction channels were taken into account. The abstractions of H_s and $H_{p,op}$ were found to be complex-forming channels, while $H_{p,ip}$ is abstracted directly. Qualitatively, the potential energy surfaces from the literature [155, 181] could be reproduced, while some small discrepancies emerge in the obtained potential energies. However, the high level calculations from this group are assumed to exhibit a high reliability.

k_3 was investigated by Kiecherer [183] in a temperature range between 250 and 750 K at pressures from 10^{-2} to 10^5 bar. For the indirect channels, chemical activation was considered and a master equation analysis was carried out. Transition state theory was applied for the bimolecular reaction steps, while the unimolecular processes with tight and loose transition states were described with RRKM theory and statistical adiabatic channel model (SACM), respectively.

In line with the findings in ref. [155], the abstraction of H_s was obtained as the main channel contributing almost solely to the overall rate coefficient at room temperature. However, with increasing temperature the direct abstraction of $H_{p,ip}$ gains in importance. At 750 K its branching ratio amounts to 45 %. The contribution of the p,op-channel is less than 5 % at all temperatures.

The calculated k_3 exhibit an s-shaped fall-off behavior typical for complex-forming reactions. The extent of the pressure dependence increases with decreasing temperature. However, the difference between the low- and high-pressure limit is relatively small amounting to maximum 50 % of the low-pressure limit. The reason for this could be found in the barriers of the forward and backward reactions of the prereactive complex. These reactions are the only unimolecular steps of R3 and thus solely exhibit a pressure dependence. Since the heights of the corresponding reaction barriers in the main channel differ only by 9.3 kJ mol^{-1}, the overall k_3 depends only slightly on the pressure, although the rate coefficients of the individual steps show a strong correlation.

With these findings, the absence of a pressure dependence of the rate coefficient of the reaction DME + OH, k_2, (cf. chapter 5 and refs. [18, 157]) could also be explained through conclusion by analogy. Here, a very similar reaction mechanism was found. However, the barrier heights of the forward and backward reaction of the prereactive complex in the main channel differ even less by 3.4 kJ mol^{-1}. Hence, it can be concluded that the low- and high-pressure limits of k_2 coincide due to these mechanistic properties.

Another conspicuity of the theoretical results is the position of the fall-off regime of k_3. It is situated at relatively high pressures of at least 10 bar. This results from low absolute values of the barriers of the unimolecular reaction steps. As a consequence, the life-time of the prereactive complex is short, ranging from 0.1 to 6 ps under the examined conditions. Hence, high pressures are needed to stabilize the short-living complex.

At lower temperatures the calculated k_3 decrease with increasing temperature passing a narrow minimum at approximately 600 K. For higher temperatures a pronounced positive temperature dependence was obtained. This behavior arises from the increasing influence of the direct p,ip-channel at higher temperatures. While the overall bimolecular rate coefficient of the s-channel decreases continuously with the temperature in the whole considered temperature range, the rate coefficient of the p,ip-channel exhibits a pronounced positive temperature dependence. Consequently, an abrupt transition from a negative to a positive temperature dependence results for the overall rate coefficient in the region where both channels contribute equally.

A comparison of the temperature dependence of k_3 obtained from the theoretical studies of this group [183] and of Zavala-Oseguera *et al.* [155] as well as the results of the direct experiments [118, 142, 144, 147] and the high temperature measurement from ref. [148] are depicted in figure 6.2. Although both theoretical studies are able to reproduce the order of magnitude of the experimentally determined k_3, some deviations in the predicted Arrhenius behavior can be observed. The discrepancies between the results of Zavala-Oseguera *et al.* and the experimental literature data have already been discussed. The theoretical study of this group [183] is at least able to reproduce the general qualitative trend of the temperature dependence.

To further validate the theoretical findings, the rate coefficient of the reaction of the perdeuterated diethyl ether DEE-d10 + OH, k_{3d}, was also calculated by Kiecherer [183]. A similar Arrhenius behavior with a minor pressure dependence of the corresponding rate coefficient k_{3d} was obtained. The isotope effect, characterized by the ratio k_3/k_{3d}, ranges from 2.0 to 2.5. Hence, it is significantly smaller than the one of the system DME + OH (cf. chapter 5). Again, the reason for this difference can be found in the barrier height of the forward reaction of the prereactive complex. This reaction step is the only step in the main channel which is significantly influenced by the perdeuteration. However, its barrier is decreased in the system DEE(-d10) + OH in comparison to the system DME(-d6) + OH by approximately 6 kJ mol^{-1}. As a result this step contributes less to the overall rate coefficient and the isotope effect of k_3 is reduced.

No experimental study of k_{3d} is available in the literature. Wallington *et al.* [144] mentioned values for k_{3d} determined by Tully in a temperature range between 296 and 441 K,

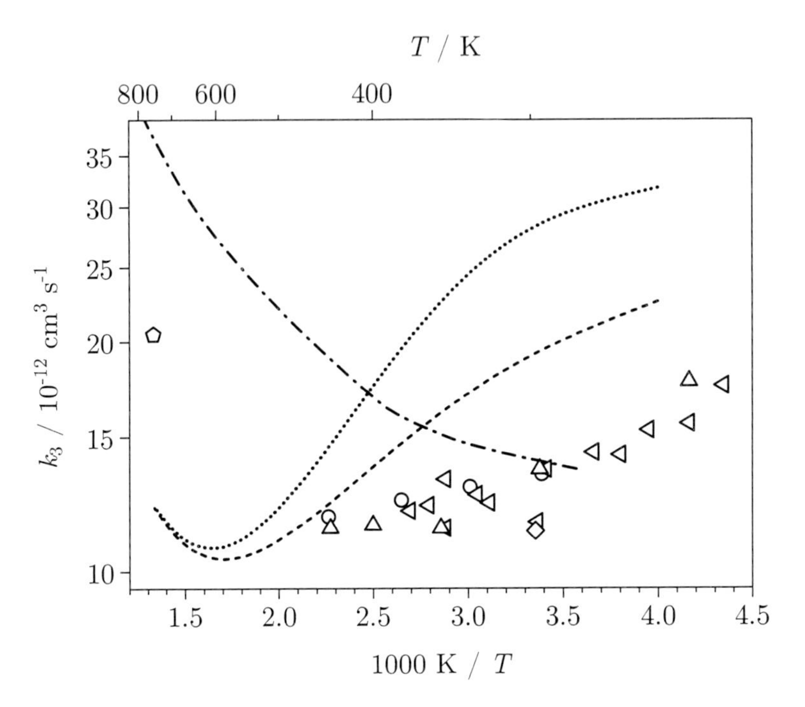

Figure 6.2: Arrhenius plot of literature data for k_3. Theoretical studies: this group, 1 bar [183] (- -), this group, high-pressure limit [183] (····), Zavala-Oseguera *et al.*, high-pressure limit [155] (— ·); Experimental studies: Tully and Droege, 0.53 bar [147] (○), Wallington *et al.*, 0.03-0.07 bar [144] (△), Nelson *et al.*, 1 bar [142] (◇), Mellouki *et al.*, 0.13 bar [118] (◁), Tranter and Walker [148], 0.67 bar (⬠).

citing them as private communication. These rate coefficients show no significant temperature dependence and indicate an isotope effect k_3/k_{3d} of 1.7 to 2.0 [144]. However, no direct publication of these values with the description of the experimental approach can be found in the literature.

Hence, the aim of this work was the experimental investigation of k_3 and k_{3d} at elevated temperatures and pressures with the PLP/LIF technique. On the one hand, the objective was to provide a reliable parametrization for k_3 under autoignition-relevant conditions, which can directly be applied in model mechanisms for the description of the ignition process of DEE. On the other hand, a validation of the theoretical results should be achieved, in order to derive generalized conclusions with respect to the mechanistic details, which are transferable on similar systems.

6.2 Experimental Procedure

The revised setup for gas injection (see section 3.1) was applied for the experimental investigation of $k_{3(\mathrm{d})}$. Experiments were carried out at temperatures between 295 and 570 K at 2, 5 and 10 bar pressure and a repetition rate of 10 Hz. Altogether, the concentrations were in the following ranges: $2.0 \cdot 10^{14}$ cm^{-3} $\geq$ [DEE] $\geq 4.13 \cdot 10^{16}$ cm^{-3}; $3.86 \cdot 10^{14}$ cm^{-3} $\geq$ [DEE-d10] $\geq 5.60 \cdot 10^{16}$ cm^{-3}; $3.86 \cdot 10^{15}$ cm^{-3} $\geq$ [HNO_3] $\geq 4.92 \cdot 10^{16}$ cm^{-3}.

With the regulation of the individual gas flows the concentrations of the different components could be controlled. In the ordinary series of measurements the overall gas flow was held constant at a value of 3 slm. The flow of the HNO_3 mixture amounted to 1 slm. The DEE(-d10) concentration was changed by varying the flow of the DEE(-d10) mixture stepwise by 0.2 slm from 0 to 1.8 slm or vice versa in each ordinary series of measurements. To balance the alteration, the flow of pure helium was changed correspondingly in a range between 0.2 and 2 slm. A potential flow and repetition rate dependence of the measured pseudo-first order rate coefficients $k'_{3(\mathrm{d})}$ was tested at 294 K and 10 bar and 570 K and 2 bar, respectively. For this, all individual gas flows were altered by a constant factor starting from initial values of 1.2 slm for the HNO_3 mixture, 1.0 slm for the DEE mixture, and 1.3 slm for pure helium. Hence, the overall gas flow was varied in a range between 0.7 and 9.1 slm. The repetition rate was reduced to 5 and 1 Hz.

The fluence of the photolysis laser varied in a range between 8 and 19 mJ cm^{-2}, while the excitation laser operated between 23 and 89 mJ cm^{-2}. A slight decalibration of the dye laser was detected during this work (cf. section 3.2). As a result it was adjusted by optimizing the OH fluorescence signal through a variation of the excitation wavelength. The measurements of $k_{3(\mathrm{d})}$ were conducted with two different settings of the dye laser wavelength of 281.923 and 281.937 nm. Different settings of the monochromator parameters were also tested. The detection wavelength amounted to 308 and 316 nm, while the slit was varied between 1 and 2 mm according to a band width between 8 and 18 nm (fwhm), respectively.

At a given time delay the integrated PMT signals of ten pulses were averaged. Approximately 90 of such averaged data points at different time delays were measured for the record of one LIF intensity-time profile. The measurement was not stopped before the PMT signal reached the zero baseline again. One series of measurements was carried out at each fixed temperature and pressure. Generally, it consisted of ten LIF intensity-time profiles at different DEE(-d10) concentrations varied in a range of one order of magnitude and one profile in absence of the reactant.

Five different DEE mixtures, three different DEE-d10 mixtures and twelve different HNO_3 mixtures were applied in the study. The age of the DEE(-d10) mixtures ranged from 1 to 97 days. Because of the slow decomposition of HNO_3 inside the gas cylinders (cf. chapter 4), the applicability of the HNO_3 mixtures was temporally limited. Consequently, the age of the HNO_3 mixtures never exceeded 12 days and the HNO_3 was always freshly prepared prior to making a mixture. The synthesis is described in section 3.1. The following chemicals were used: DEE $\geq$ 99.7 %, Sigma-Aldrich; DEE-d10 99 %, Deutero GmbH; H_2SO_4 98 %, Roth; KNO_3 $\geq$ 99 %, Roth; He > 99.999 %, Air Liquide.

6.3 Analysis

Representative LIF intensity-time profiles of one series of measurements are shown in figure 6.3.

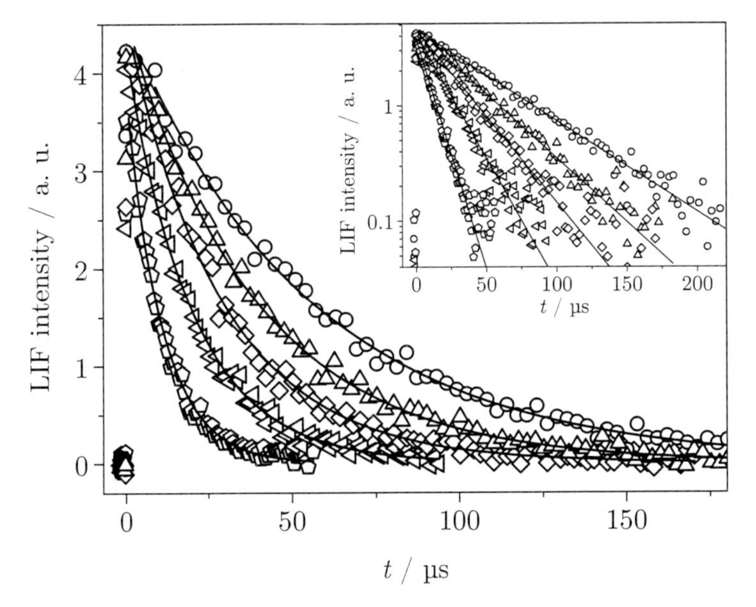

Figure 6.3: LIF intensity-time profiles of the reaction DEE + OH at 5 bar, 426 K, and different [DEE]: $0.70 \cdot 10^{15}$ cm^{-3} (○), $1.29 \cdot 10^{15}$ cm^{-3} (△), $1.92 \cdot 10^{15}$ cm^{-3} (◇), $3.14 \cdot 10^{15}$ cm^{-3} (◁), $6.27 \cdot 10^{15}$ cm^{-3} (⬠). For a better illustration the linearized signals are shown as well.

All obtained profiles could be fitted adequately with a monoexponential decay function given by

$$I = I_0 \exp(-k'_{3(\mathrm{d})} t)\,. \tag{6.1}$$

The fitting was conducted with the help of a LabView program written by Hetzler [184] in this group. Here, the zero baseline is calculated automatically by averaging the measuring points recorded at a negative time delay between photolysis and detection. Choosing the limits by hand, the program determines the best fit with the least-squares procedure.

All obtained pseudo-first order rate coefficients $k'_{3(\mathrm{d})}$ and the corresponding errors and measuring conditions are listed in appendix A.3. The $k'_{3(\mathrm{d})}$ were plotted against the reactant

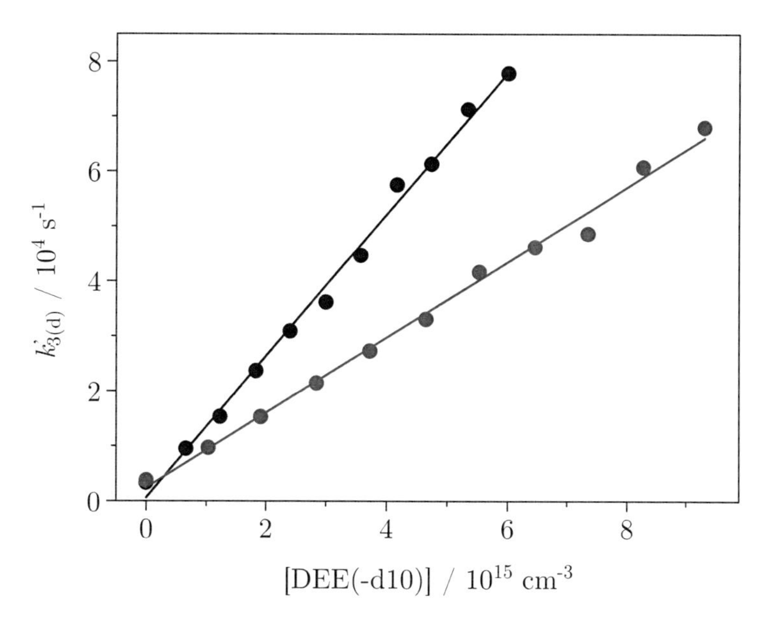

Figure 6.4: Plot of k'_3 (●, 357 K) and k'_{3d} (●, 355 K) against [DEE] and [DEE-d10], respectively, both at 2 bar pressure.

concentration [DEE(-d10)], which was obtained from the following expression:

$$[\text{DEE(-d10)}] = \frac{P_{\text{DEE(-d10),mix}}}{P_{\text{mix}}} \frac{f_{\text{DEE(-d10)}}}{f_{\text{DEE(-d10)}} + f_{\text{HNO3}} + f_{\text{He}}} \frac{P}{\text{RT}} \text{N}_\text{A} \,. \tag{6.2}$$

Here, $P_{\text{DEE(-d10),mix}}$ is the initial partial pressure of DEE(-d10) and P_{mix} the initial overall pressure of the reactant gas mixture, while P stands for the overall pressure in the reaction cell. $f_{\text{DEE(-d10)}}$, f_{HNO3}, and f_{He} represent the gas flows of the DEE(-d10) mixture, the HNO_3 mixture, and the pure helium, respectively. For all gases an ideal behavior is assumed.

Typical examples of plots of $k'_{3(\text{d})}$ against [DEE(-d10)] are depicted in figure 6.4. The bimolecular rate coefficients $k_{3(\text{d})}$ were obtained from the best linear fit according to

$$k'_{3(\text{d})} = k'_{\text{SR}} + k_{3(\text{d})}[\text{DEE(-d10)}] \,. \tag{6.3}$$

The intercepts in the $k'_{3(\text{d})}$-plots represent the sum of the pseudo-first order rate coefficients of the side reactions k'_{SR} (cf. chapter 4). For all series of measurements it amounted to less than 36 % of the corresponding maximum $k'_{3(\text{d})}$ at the highest reactant concentration.

For all series of measurements the assumption of a linear relation between $k'_{3(\text{d})}$ and [DEE(-d10)] was adequate. In addition, the validity of pseudo-first order conditions was verified by estimating the initial OH concentrations according to equations 3.2 and 3.1. Values between $7.31 \cdot 10^{11}$ and $2.08 \cdot 10^{13}$ cm^{-3} resulted for the upper limit of $[\text{OH}]_0$. Hence, the $[OH]_0/[DEE(\text{-d10})]$-ratio never exceeded 1 %. Hence, pseudo-first order conditions are assumed to be valid under the present conditions.

To eliminate various sources of error, the correlation between several parameters were tested (cf. subsection 3.2.5). As described in section 6.2 a possible dependence of the pseudo-first order rate coefficients $k'_{3(d)}$ of the overall gas flow and the repetition rate was tested in multiple series of measurements. It could be excluded under all conditions. Moreover, a correlation between the bimolecular rate coefficients and the age of the gas mixtures as well as the laser fluences was surveyed. Here, the difference between the measured $k_{3(d)}$ and the corresponding value of the Arrhenius fit (cf. section 6.4) was plotted as a function of the particular parameter. No systematic dependence was observed in all cases.

6.4 Results and Discussion

With each series of measurements one bimolecular rate coefficient at a defined temperature and pressure was determined. A potential pressure dependence of $k_{3(d)}$ was examined for all considered temperatures. Figure 6.5 shows a plot of the rate coefficients k_3 at room temperature against the pressure comparing the results from this work and the direct studies from the literature (cf. section 6.1). Moreover, the k_3 calculated by Kiecherer [183] are depicted.

The experimental values agree well within a range of 17 %. No correlation with the pressure emerges under the present conditions. Although the calculated k_3 overestimate the experimental values by approximately 50 %, the absence of a pressure dependence is well predicted. According to the theory, a significant increase of k_3 at room temperature can only be observed for pressures higher than 20 bar. Hence, it can be assumed that the experimentally determined values characterize the low-pressure limit.

Generally, for all temperatures no systematic dependence of $k_{3(d)}$ on the pressure was observed in this work. As a result the $k_{3(d)}$ obtained at one particular temperature and 2, 5, and 10 bar pressure were averaged. The resulting mean values are listed in table 6.1 for the system DEE + OH and in table 6.2 for the system DEE-d10 + OH.

The rate coefficients k_3 obtained in this work showed a negligible temperature dependence. Hence, the values were fitted with a simple Arrhenius expression according to

$$k_3 = 1.27 \cdot 10^{-11} \exp\left(\frac{4\ \mathrm{K}}{T}\right)\ \mathrm{cm^3\ s^{-1}}\,. \tag{6.4}$$

In Figure 6.6 the averaged values for k_3 are compared to all experiments from the literature in an Arrhenius plot. It illustrates the disagreement between the results of the direct experiments [118, 142, 144, 147] and some competitive relative rate studies [177, 179, 180], as already discussed in section 6.1.

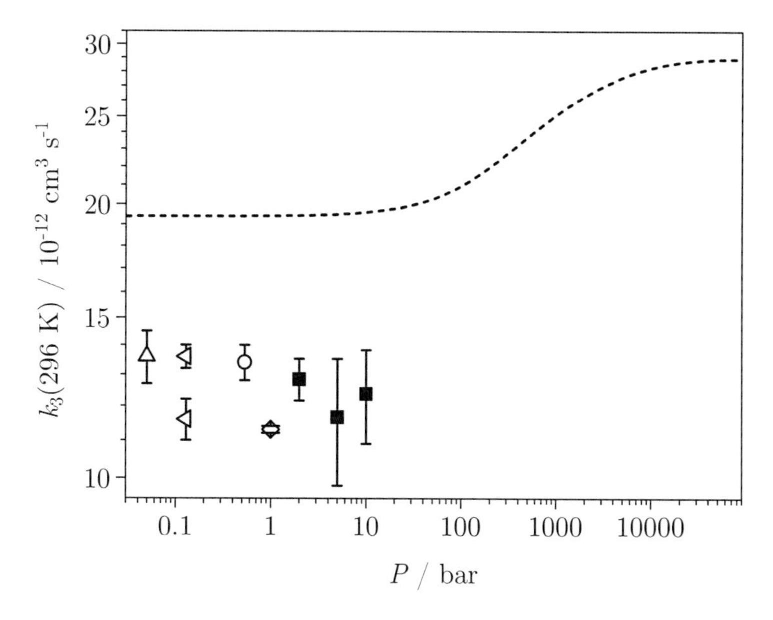

Figure 6.5: Plot of k_3 at room temperature against the pressure: Experiments: this work (■); Tully and Droege [147] (O); Wallington *et al.* [144] (△); Nelson *et al.* [142] (◇); Mellouki *et al.* [118] (◁); Theory: Kiecherer [183] (- -).

However, taking into account only the direct studies of k_3 from the literature and of this work, a reasonably good agreement emerges. It is illustrated in figure 6.7. A curved Arrhenius behavior with a minimum at around 400 K arises. Thus, k_3 was parametrized with a modified Arrhenius equation including all values for k_3, which are depicted in figure 6.7. The following expression was obtained as best fit:

$$k_3 = 1.46 \cdot 10^{-17} \left(\frac{T}{\mathrm{K}}\right)^{1.948} \exp\left(\frac{778\ \mathrm{K}}{T}\right)\ \mathrm{cm}^3\ \mathrm{s}^{-1}. \tag{6.5}$$

With a deviation of approximately 20 %, the extrapolation of this relation to higher temperatures is in good agreement with the measurement of Tranter and Walker [148] at 753 K. However, in order to examine the slope of the positive temperature dependence more precisely, further validation at high temperatures is recommendable.

In figure 6.7 the results for the rate coefficient $k_{3\mathrm{d}}$ of the system DEE-d10 + OH from this work and from Tully mentioned by Wallington *et al.* in ref. [144] are also presented. Both studies are in a very good agreement indicating a slightly positive temperature dependence with a small curvature in the Arrhenius behavior. However, the data from ref. [144] are cited as private communication. No direct publication of the study with a description of the experimental approach could be found in the literature.

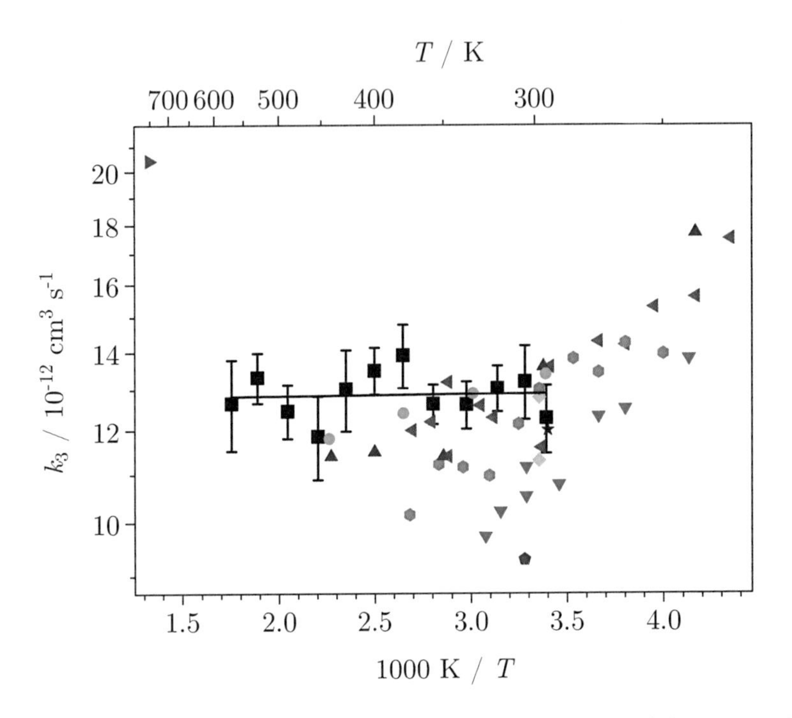

Figure 6.6: Arrhenius plot of experimentally determined k_3: this work (■), best fit of a simple Arrhenius expression according to equation 6.4 (—); Tully and Droege [147] (●); Wallington *et al.* [144] (▲); Nelson *et al.* [142] (◆); Mellouki *et al.* [118] (◀); Lloyd *et al.* [177] (⬟); Bennett and Kerr [178] (★); Bennett and Kerr [179] (▼); Semadeni *et al.* [180] (⬢); Tranter and Walker [148] (▶).

T / K	$\overline{k_3}$ / 10^{-12} cm^3 s^{-1}
295 ± 1	12.3 ± 0.8
305 ± 1	13.2 ± 1.0
319 ± 2	13.0 ± 0.6
336 ± 2	12.6 ± 0.6
357 ± 1	12.6 ± 0.5
378 ± 1	13.9 ± 0.9
400 ± 1	13.5 ± 0.6
426 ± 4	13.0 ± 1.0
454 ± 1	11.9 ± 1.0
489 ± 1	12.5 ± 0.7
529 ± 1	13.3 ± 0.7
570 ± 1	12.7 ± 1.1

Table 6.1: Overview of all averaged rate coefficients $\overline{k_3}$ obtained in this work.

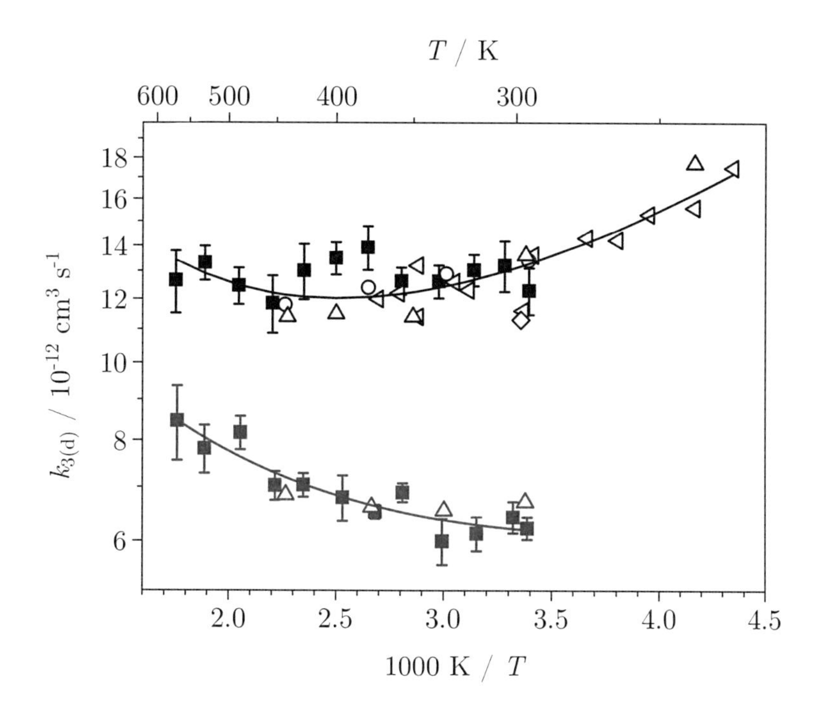

Figure 6.7: Arrhenius plot of k_3 (black) and k_{3d} (red) from direct experiments: this work (■); Tully and Droege [147] (○); Wallington *et al.* [144] (△); Nelson *et al.* [142] (◇); Mellouki *et al.* [118] (◁); Arrhenius fits according to equations 6.5 and 6.6 (—).

T / K	$\overline{k_{3d}}$ / 10^{-12} cm^3 s^{-1}
295 ± 1	6.2 ± 0.2
301 ± 1	6.4 ± 0.3
317 ± 2	6.1 ± 0.3
334 ± 1	6.0 ± 0.4
356 ± 1	6.9 ± 0.2
373 ± 1	6.52 ± 0.13
395 ± 1	6.8 ± 0.4
426 ± 5	7.0 ± 0.2
451 ± 3	7.0 ± 0.3
486 ± 2	8.2 ± 0.4
529 ± 1	7.8 ± 0.5
568 ± 1	8.5 ± 0.9

Table 6.2: Overview of all averaged rate coefficients $\overline{k_{3d}}$ obtained in this work.

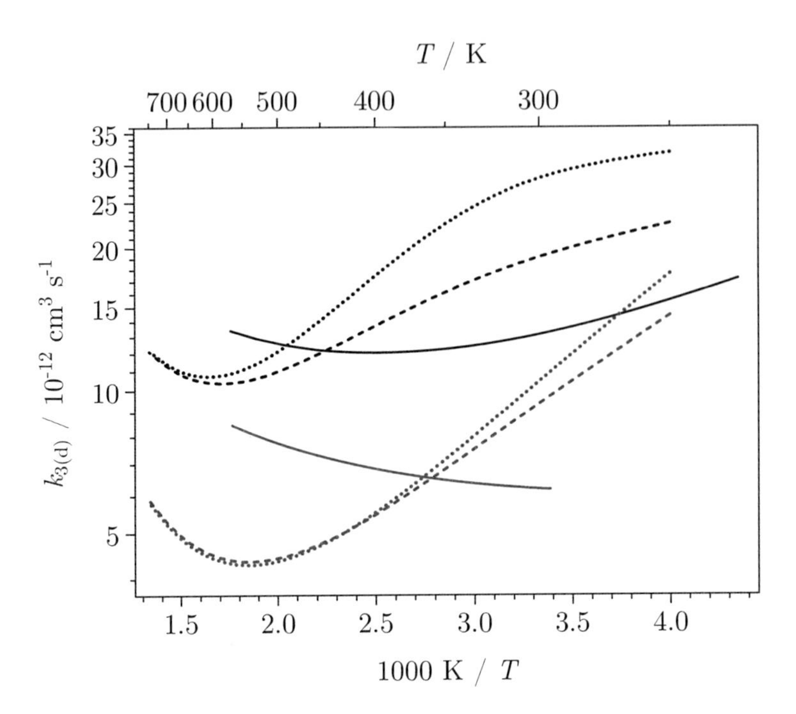

Figure 6.8: Arrhenius plot of k_3 (black) and k_{3d} (red): Arrhenius fits to experimental values according to equations 6.5 and 6.6 (—); Calculation [183] at 1 bar (- -); Calculation [183] at the high-pressure limit (····).

Hence, only the values of the present work were included in the parametrization described by the following expression:

$$k_{3d} = 9.82 \cdot 10^{-16} \left(\frac{T}{\mathrm{K}}\right)^{1.334} \exp\left(\frac{343\ \mathrm{K}}{T}\right)\ \mathrm{cm^3\ s^{-1}}\,. \tag{6.6}$$

In Figure 6.8, a comparison between the Arrhenius fits from equations 6.5 and 6.6 and the calculations of Kiecherer [183] is given. For the undeuterated case the experimental findings can be qualitatively reproduced, although the theoretically predicted minimum is a bit more narrow and slightly shifted to higher temperatures. In the case of the perdeuterated system the deviations are more pronounced. While the experiments yielded a positive temperature dependence in the whole temperature range, a negative correlation was obtained from the theory under the corresponding conditions.

Nevertheless, the isotope effect of the overall rate coefficients k_3/k_{3d} is well reproduced by the theory. In Table 6.3 the experimental and theoretical values at different temperatures are compared. Taking into account that *ab initio* calculations were applied without any adjustments, the agreement is remarkable and further validates the theoretical findings.

method	k_3/k_{3d}					
	300 K	350 K	400 K	450 K	500 K	550 K
exp. this work	2.1	1.9	1.8	1.7	1.6	1.6
theory [183]	2.1	2.4	2.5	2.5	2.5	2.4

Table 6.3: Isotope effect k_3/k_{3d} from experiment and theory at different temperatures.

6.5 Conclusion and Outlook

The range of conditions at which k_3 is determined experimentally could be expanded up to 570 K and 10 bar by the measurements in this work. With the good agreement between all direct studies of k_3, an overall parametrization of the Arrhenius behavior between 230 and 570 K could be achieved. No pressure dependence was observed between 0.05 and 10 bar. With the help of the theoretical investigation carried out within this group [183] this finding could be explained in terms of the short lifetime of the prereactive complex in the complex-forming main channel. As a result, the fall-off region of k_3 is situated at pressures higher than 20 bar and the Arrhenius expression obtained from the experiments describes the low-pressure limiting rate coefficient. However, due to the generally small extent of the pressure dependence of k_3, especially at ignition-relevant temperatures, it is assumed to be negligible for the description of combustion processes of DEE. Hence, the presented parametrization is recommended to be applied in model mechanisms for DEE ignition and combustion without consideration of a pressure dependence even if higher pressures are regarded.

With respect to temperatures above 600 K, further experimental validation of the obtained parametrization would be valuable. The results of this work in combination with the direct studies from the literature seem to characterize the minimum in the Arrhenius plot well. However, the following increase of k_3 at higher temperatures can only be speculated. In addition, the theoretical description of k_3 does not exhibit the sufficient accuracy for a direct application of the absolute values. Consequently, only an experimental investigation of the Arrhenius behavior in the high temperature region supplementary to the study of Tranter and Walker [148], could identify the slope of the positive temperature dependence in this regime more accurately.

Despite the limited agreement between the absolute theoretical values for k_3 from Kiecherer [183] and the experiment, the general implications of the theoretical findings could be validated by means of several experimental observations. First of all, the absence of a pressure dependence in the experimental studies of k_3 as well as of k_2 (cf. chapter 5 and refs. [18, 157]) confirms the theoretically described fall-off behavior. Second, the agreement in the qualitative trend of the Arrhenius behavior of k_3 obtained by experiment and theory substantiates the various influences of the particular reaction channels at different temperatures predicted by the theory. Last, the theoretically determined isotope effect of k_3 is in good agreement with the experimental findings. Basically, this supports the validity of the computed potential energy surface.

7 The Reaction DMM + OH

7.1 Introduction

Polyoxymethylene dimethyl ethers (POMDMEs or OMEs) are a class of polyethers with the general formula $H_3CO(CH_2O)_nCH_3$. The first homolog in this series with $n = 1$ is the smallest diether, dimethoxymethane (DMM), also referred to as methylal. It is applied as a solvent with minor importance [185], while the higher homologs are of no industrial relevance to date. Hence, only small amounts of DMM are produced at the moment [186]. However, in the case of an increased demand, different synthesis routes on the basis of methanol are available and seem to make a cost-efficient large scale production of POMDMEs from fossil feedstock as well as from biomass possible. [16, 187]

In the search for specifically adapted, emission-reduced designer biofuels (cf. section 5.1 and 6.1) POMDMEs have recently come into focus. They hold advantages similar to DME like the relative oxygen content, the absence of any carbon-carbon bonds and low ignition temperature. Beyond that, POMDMEs exhibit significantly improved material properties in comparison to DME. They are, for example, liquid under standard conditions. As a result, DMM and the higher POMDMEs up to $n = 6$ are considered as potential diesel fuel additives and substitutes. In fact, first studies on the combustion and ignition properties gave promising results with respect to pollutant emission and performance characteristics. [16, 186, 188]

Several DMM oxidation mechanisms have already been published and modified [189–194]. They are not only relevant for DMM combustion itself, but also serve as a basis mechanism for combustion modeling of the higher POMDMEs [195]. However, the DMM elementary reactions which are involved are only poorly investigated as yet. As a result, all rate coefficients of those reactions were estimated mainly on the basis of DME, DEE, and propane reactions in the mechanisms of refs. [189, 191, 192].

The reaction of a hydroxyl radical with DMM

$$\mathrm{DMM} + \mathrm{OH} \xrightarrow{k_4} \mathrm{products} \tag{R4}$$

is an important elementary process in DMM combustion and crucial for the modeling of the autoignition of DMM. In the case of a future application as a fuel blend and hence a more

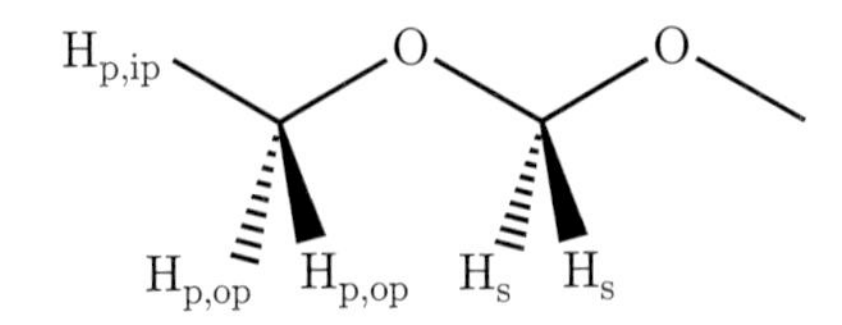

Figure 7.1: Structure of dimethoxymethane (DMM) with labeling of the different types of H atoms.

extensive emission of DMM itself, this reaction would also become important in the context of DMM degradation in the atmosphere [93, 196].

Figure 7.1 illustrates the differentiation between the individual types of H atoms, which can be abstracted in reaction R4: the primary H atoms of the methyl groups in out-of-plane position $H_{p,op}$ and in-plane-position $H_{p,ip}$, and the secondary H atoms H_s at the methylene site.

Due to the lack of direct literature data which had been verified under combustion conditions, Daly *et al.* [189] estimated the rate coefficient k_4. They distinguished between the abstraction of primary and secondary H atoms. The rate coefficient of the first channel was estimated with the help of one for the reaction DME + OH of Curran *et al.* [197]. This value in turn was an estimation based on the measured rate coefficient for propane + OH from ref. [198]. The kinetics of the second channel was described with a calculated rate coefficient of the reaction propane + OH from Cohen [199]. In the mechanism of Dias *et al.* [191] an experimentally determined rate coefficient of the reaction DME + OH of DeMore and Bayes [146] was applied for k_4. Marrodán *et al.* [192] estimated k_4 on the basis of DME + OH as well. However, they employed the parametrization of Arif *et al.* [95] because of the broader temperature range of the determination. Nevertheless, all estimations applied in the different mechanisms agree quite well with each other. They show a pronounced positive temperature dependence with a slight curvature from room temperature up to combustion-relevant conditions. A possible pressure dependence of k_4 was neglected in all mechanisms.

No theoretical study on the reaction R4 is published yet. However, there are four different experimental studies available in the literature dealing with a direct investigation of k_4 [114, 200–202]. Porter *et al.* [200] applied two different measuring methods. On the one hand, they carried out experiments with pulse laser photolysis/resonance fluorescence in a temperature range between 230 and 372 K at 133 mbar. On the other hand, relative rate measurements were conducted at 298 K and 1 bar. The results of both techniques agree well with each other. Thus, no dependence of k_4 on the pressure was deduced. The observed temperature dependence is slightly negative with a small curvature. Wallington *et al.* [201] applied a relative rate method and pulse radiolysis/transient UV absorption technique for their measurements at 295 K and 346 K, respectively, both at a pressure of 1 bar. In good agreement with the findings of Porter *et al.* [200], they obtained a slight decrease of k_4 with temperature. A relative rate method was also used by Thüner *et al.* [202] for their experiments at 298 K and 1 bar. The results of the former studies at room temperature could be confirmed. The most recent work of Vovelle *et al.* [114] with pulsed laser photolysis/laser-induced fluorescence (PLP/LIF) technique is the only study at autoignition-relevant temperatures from 293 to 617 K. The pressures amounted to 67 and 133 mbar. Contrary to the observations of Porter *et al.* [200] and Wallington *et al.* [201], they obtained a pronounced positive temperature dependence over the whole temperature range of their experiments. No dependence on the pressure was observed. Despite the discrepancy in the temperature dependence, the agreement of the absolute rate coefficients with the other studies in the overlapping temperature region lies in the error range of the experiment. Hence, a strongly curved Arrhenius behavior with a minimum at around 330 K was assumed and a parametrization with a modified Arrhenius expression including all literature values was achieved.

In the present work the kinetics of reaction R4 was studied with pulse laser photolysis/laser-induced fluorescence (PLP/LIF) under autoignition-relevant temperatures and pressures. The aim was to verify the previous experimental findings with respect to the pressure and temperature dependence and to provide an experimentally determined direct parametrization of k_4 for the modeling of DMM combustion. The experimental results were interpreted with regard to the characteristics of the potential energy surface on the basis of the theoretical findings in the studies of the systems DME(-d6) + OH and DEE(-d10) + OH (see refs. [18, 157, 183], cf. chapters 5 and 6).

7.2 Experimental Procedure

The measurements of the rate coefficient k_4 were realized with the revised setup for gas injection, which is described in section 3.1. Experiments were carried out at 2, 5, and 10 bar pressure and twelve different temperatures in a range from 297 to 570 K. A repetition rate of 10 Hz was chosen. Altogether, [DMM] ranged from $6.2 \cdot 10^{14}$ to $5.3 \cdot 10^{16}$ cm^{-3} and [HNO_3] from $5.2 \cdot 10^{15}$ to $7.8 \cdot 10^{16}$ cm^{-3}.

The concentrations were regulated by adjusting the gas flows of the different flow controllers. In an ordinary series of measurements the overall gas flow and the flow of the HNO_3 mixture were set to 3.5 slm and 1.2 slm, respectively. The flows of the DMM mixture and pure helium were varied in steps of 0.2 slm from 0.0 to 2.0 slm and 2.3 to 0.3 slm, respectively, or vice versa. In two extra series of measurements, one at 297 K and 10 bar and one at 566 K and 2 bar, a potential dependence of the measured pseudo-first order rate coefficient on the overall gas flow and the repetition rate of the experiment (cf. subsection 3.2.5) was examined. Starting from 1.2 slm for the HNO_3 mixture, 1.0 slm for the DMM mixture, and 1.3 slm for pure helium, all flows were varied by a factor ranging from 0.2 to 2.6. With this procedure the concentrations were held constant, while the overall gas flow ranged from 0.7 to 9.1 slm. Moreover, two measurements were carried out at a reduced repetition rate of 1 and 5 Hz.

The laser fluence of the photolysis laser ranged from 6 to 17 mJ cm^{-2} and the fluence of the excitation laser from 19 to 61 mJ cm^{-2}. The problem of a slight decalibration of the dye laser wavelength is discussed in section 3.2. Before starting the measurements the detection wavelength was adjusted by optimizing the observed OH-LIF signal at fixed conditions. At 281.937 nm the most intense signal was detected. For all measurements of the system DMM + OH the detection wavelength was set to this value. The monochromator wavelength was tuned to 308 nm and the slit to 1 mm. This corresponds to a band width of 8 nm (fwhm).

At each fixed temperature and pressure, one series of measurements was carried out consisting of eleven LIF intensity-time profiles. Ten of them were recorded at different DMM concentrations varied in a range of approx. one order of magnitude. Additionally, one profile was measured without the reactant DMM. For each LIF intensity-time profile approximately 90 measuring points were recorded. Care was taken that the PMT signal reached the zero baseline again before the recording was stopped. Each measuring point at a given time delay was the mean value of 10 single points averaged internally in the measuring program.

Two different DMM mixtures and five different HNO_3 mixtures were used. The age of the applied DMM mixture ranged from 3 to 29 days. In contrast, the age of the HNO_3 mixtures was purposely held low because of the slow decomposition of HNO_3 inside the gas cylinders (cf. chapter 4). It never exceeded four days. Additionally, HNO_3 was synthesized directly before an HNO_3 gas mixture was prepared following the procedure described in section 3.1. All substances were degassed carefully before making a mixture. The following chemicals were employed: DMM $\geq$ 99.0 %, Sigma-Aldrich; H_2SO_4 98 %, Roth; KNO_3 $\geq$ 99 %, Roth; He $>$ 99.999 %, Air Liquide.

7.3 Analysis

Representative LIF intensity-time profiles of one series of measurements and their best fits are shown in figure 7.2.

All measured LIF intensity-time profiles of the system DMM + OH showed a mono-exponential decay. Thus, with the best fit of the expression

$$I = I_0 \exp\left(-k_4' t\right) \tag{7.1}$$

k_4' was obtained. The fitting was performed with a LabView program written by Hetzler [184] in this group (cf. section 6.3).

All determined k_4' with the corresponding measuring conditions are listed in appendix

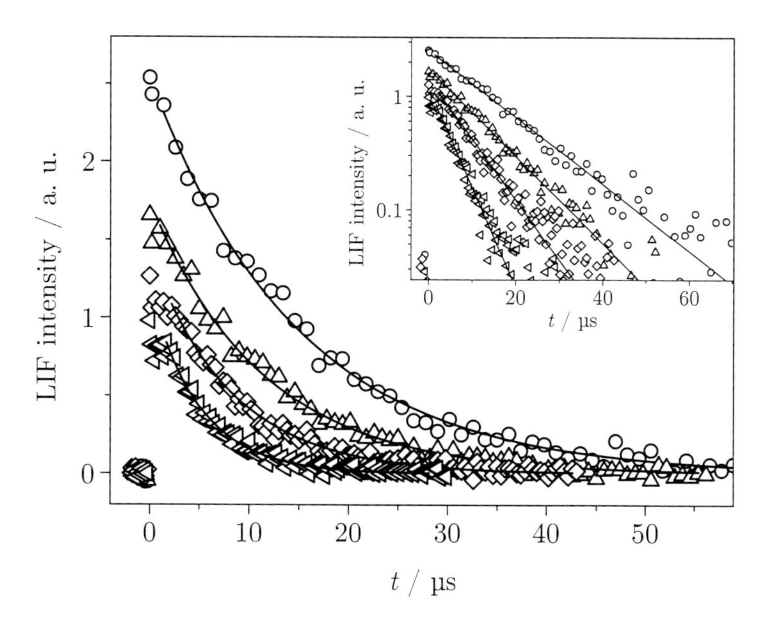

Figure 7.2: LIF intensity-time profiles of the reaction DMM + OH at 10 bar, 529 K, and different [DMM]: $3.33{\cdot}10^{15}$ cm^{-3} (○), $6.66{\cdot}10^{15}$ cm^{-3} (△), $1.67{\cdot}10^{16}$ cm^{-3} (◇), and $3.33{\cdot}10^{16}$ cm^{-3} (◁). For a better illustration the linearized signals are shown as well.

A.4. The values of k_4' were plotted against the reactant concentration [DMM]. The latter was calculated from the initial partial pressure $P_{\mathrm{DMM,mix}}$ of DMM, the initial overall pressure P_{mix} of the reactant gas mixture, and the overall pressure in the reaction cell P under the assumption of ideal gases. The relation is described by

$$[\mathrm{DMM}] = \frac{P_{\mathrm{DMM,mix}}}{P_{\mathrm{mix}}} \frac{f_{\mathrm{DMM}}}{f_{\mathrm{DMM}} + f_{\mathrm{HNO3}} + f_{\mathrm{He}}} \frac{P}{\mathrm{R}T} \mathrm{N_A}\,, \tag{7.2}$$

where f_{DMM}, f_{HNO3}, and f_{He} are the gas flows of the DMM mixture, the HNO_3 mixture, and pure helium, respectively.

The best fit of

$$k_4' = k_{\mathrm{SR}}' + k_4[\mathrm{DMM}] \tag{7.3}$$

gave the bimolecular rate coefficient k_4 as the slope. The intercept k_{SR}' is the pseudo-first order rate coefficient of the side reactions (cf. chapter 4). It lay at maximum at 27 % of the largest k_4' of the corresponding series of measurements. Figure 7.3 shows two plots of k_4' against [DMM], which belong to the series of measurements at 297 and 570 K, both at 5 bar pressure.

No systematic deviation from the linear relation between [DMM] and k_4' was observed in any series of measurements. Additionally, the upper limit of the OH radical concentration

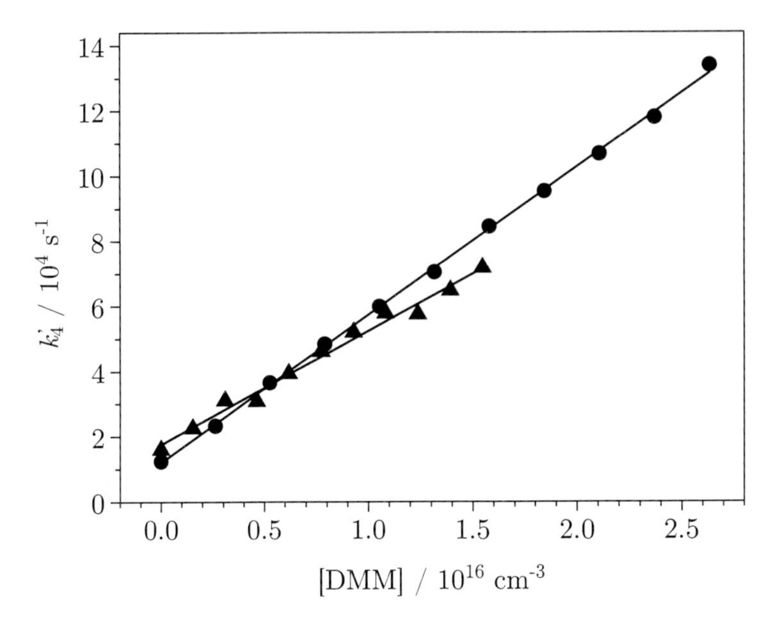

Figure 7.3: Plot of $k_4^{'}$ against [DMM] of two series of measurements at 297 K (●) and 570 K (▲), both at 5 bar pressure.

was estimated with equations 3.2 and 3.1. It ranged from $1.3 \cdot 10^{12}$ to $1.6 \cdot 10^{13}$ cm^{-3}. Hence, the ratio [OH]/[DMM] amounted to less than 0.52 %. Consequently, the assumption of pseudo-first order conditions are proven to be valid for the measurements of k_4 in this work.

According to the error analysis in section 3.2, several correlations between the measured rate coefficient and experiment-specific parameters were tested. First of all, $k_4^{'}$ was measured as a function of the flow rate and the repetition rate of the experiment in multiple series of measurements, as described in section 7.2. No systematic dependence arose. Furthermore, a potential dependence of the measured rate coefficients on the age of the gas mixtures, as well as on the laser fluences, was checked. For this purpose, the difference between the measured k_4 and the corresponding value of the Arrhenius fit (cf. section 7.4) was plotted against each parameter, respectively. In all cases, no correlation was obtained.

7.4 Results and Discussion

One bimolecular rate coefficient was obtained from each series of measurements at a certain temperature and pressure. In figure 7.4, k_4 is plotted against the reverse temperature for different pressures. At every pressure a slightly negative temperature dependence without any significant curvature in the Arrhenius behavior was observed. As a result, a simple Arrhenius equation was used for parametrization. The expressions of the best fits can be found in appendix A.4.

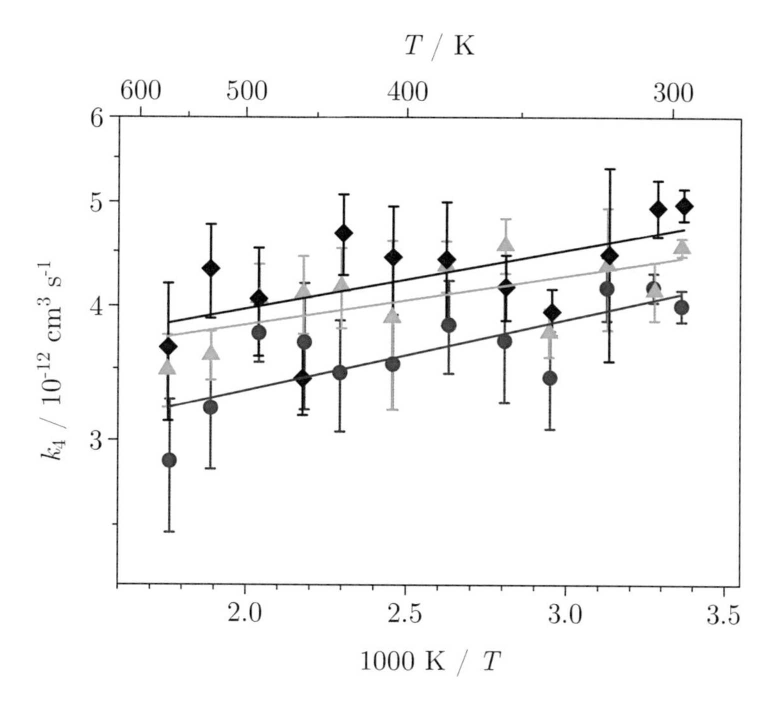

Figure 7.4: Arrhenius plot of k_4 determined in this work at 2 bar (●), 5 bar (▲), and 10 bar (◆) pressure and best fits of a simple Arrhenius expression (solid lines).

Moreover, k_4 appears to be slightly dependent on the pressure. Considering the error range and the scattering of the measurements though, the experiment cannot resolve such small variations reliably (cf. subsection 3.2.5). In figure 7.5 all values from the literature and the present work at room temperature are plotted against the pressure. Within a range of 25 % all rate coefficients agree with each other and show no correlation. Thus, no pressure dependence of k_4 can be deduced in this work.

As a result, all rate coefficients obtained at each temperature were averaged, respectively. The errors of the single values were propagated to obtain the error of the average. The resulting rate coefficients $\overline{k_4}$ are listed in table 7.1. The best fit of a simple Arrhenius expression gave:

$$k_4 = 2.90 \cdot 10^{-12} \exp\left(\frac{126\ \mathrm{K}}{T}\right)\ \mathrm{cm^3\ s^{-1}}\,. \tag{7.4}$$

Figure 7.6 shows the Arrhenius plot of all experimentally determined k_4 from the literature [114, 200–202] compared with the results of the present work. The estimation for k_4 of Marrodán *et al.* [192] is also depicted. They employed the modified Arrhenius fit to the experimentally determined rate coefficients of the reaction DME + OH of Arif *et al.* [95]. The agreement between all literature values [114, 200–202] and the results of the present work is good at moderate temperatures up to 370 K.

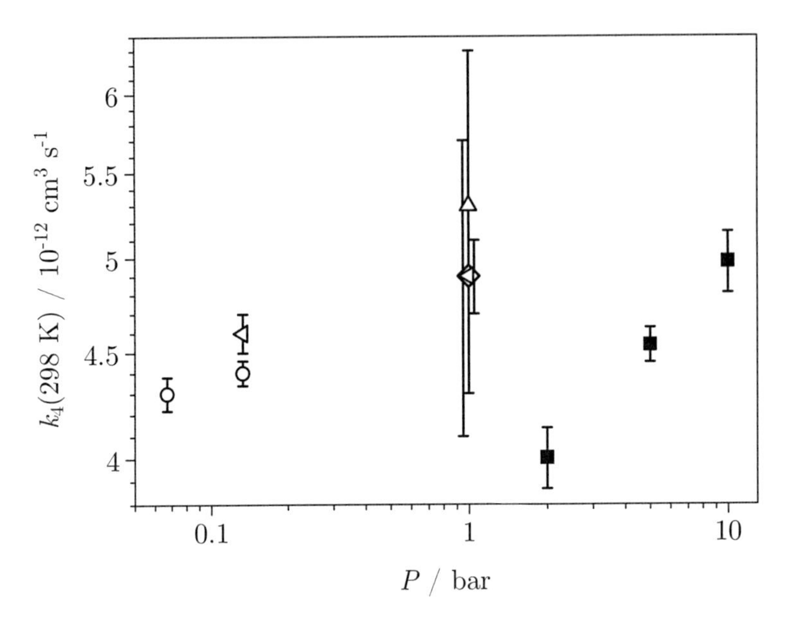

Figure 7.5: Plot of k_4 at room temperature against the pressure: this work (■); Porter *et al.* [200] (◁); Wallington *et al.* [201] (△); Thüner *et al.* [202] (◇); Vovelle *et al.* [114] (○).

T / K	$\overline{k_4}$ / 10^{-12} cm^3 s^{-1}
297 ± 1	4.5 ± 0.2
305 ± 1	4.4 ± 0.3
320 ± 1	4.3 ± 0.7
339 ± 1	3.7 ± 0.3
356 ± 1	4.1 ± 0.4
381 ± 2	4.2 ± 0.5
407 ± 1	4.0 ± 0.6
435 ± 2	4.1 ± 0.5
459 ± 1	3.7 ± 0.4
490 ± 1	4.0 ± 0.4
529 ± 1	3.7 ± 0.4
569 ± 3	3.3 ± 0.5

Table 7.1: Overview of all averaged rate coefficients $\overline{k_4}$ obtained in this work.

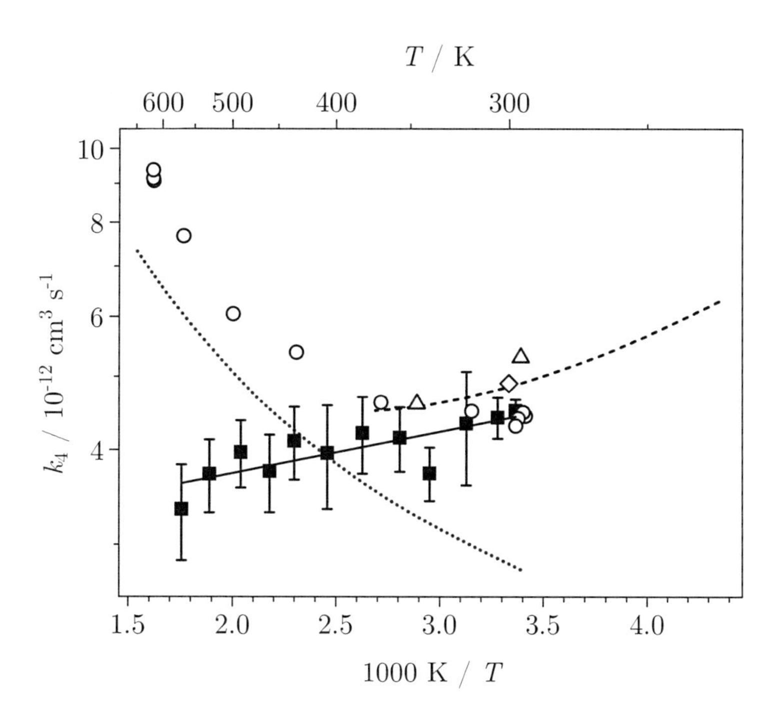

Figure 7.6: Arrhenius plot of k_4: this work (■) with best fit of a simple Arrhenius expression according to equation 7.4 (—); Porter *et al.* [200] (--); Wallington *et al.* [201] (△); Thüner *et al.* [202] (◇); Vovelle *et al.* [114] (○); estimation of Marrodán *et al.* [192] (····).

At higher temperatures, measurements were only carried out by Vovelle *et al.* [114] and in the present work. Above 400 K the results of these studies diverge considerably with increasing temperature. Vovelle *et al.* [114] found a pronounced increase of k_4 with temperature, whereas the findings of the present work indicate a continuation of the slightly negative temperature dependence obtained by Porter *et al.* [200] and Wallington *et al.* [201]. In this work no explanation could be found for these disagreeing results.

The comparison of the estimation for k_4 of Marrodán *et al.* [192] with the measurements of Vovelle *et al.* [114] shows a good agreement of the predicted temperature dependence at autoignition-relevant temperatures. The absolute values differ only by a factor of approximately 1.2. Thus, the application of this estimate is appropriate if the results of Vovelle *et al.* can be verified.

However, in the case of a confirmation of the negative temperature dependence obtained in this work, the estimation of Marrodán *et al.* [192] deviates significantly from the experimental findings. With its pronounced positive temperature dependence, there is a strong divergence with increasing temperature. At 570 K k_4 is already overestimated by a factor of 1.7. Such discrepancies can have a significant impact on the modeling results of combustion processes. Particularly for the calculation of ignition delay times, non-negligible errors can arise as the reaction of the fuel with OH radicals is crucial for the autoignition process. Consequently, in the case of a verification of the results for k_4 obtained in this work, a revision of the kinetic parameters applied for reaction R4 in the oxidation mechanism of Marrodán *et al.* [192] is recommendable.

The estimations applied in the other available mechanisms [189,191] are taken from other publications [146,197–199] (cf. section 7.1). Nevertheless, they agree very well with the one of Marrodán *et al.* [192]. As a result, the previous discussion is also valid for these models.

With the extensive studies of the systems DME + OH and DEE + OH in the present work (cf. chapters 5 and 6, respectively) and in refs. [18,157,183], a general knowledge about the reaction of ethers with OH radicals was gained. Through conclusions by analogy on the basis of these findings, fundamental information about the reaction mechanism of R4 and the rate coefficient k_4 can be derived. All theoretical findings concerning the reactions DME + OH and DEE + OH to which the following discussion is referring are entirely taken from the calculations of Kiecherer and Szöri of this group (see refs. [18,157] for DME + OH and refs. [18,182,183] for DEE + OH).

At least three different reaction channels are expected to be important in the reaction of DMM with OH radicals: the abstraction of $H_{p,op}$, $H_{p,ip}$ and H_s (cf. figure 7.1). For the abstraction of the primary H atoms, it is reasonable to assume a potential energy surface analog to the system DME + OH. The potential energy diagram of this reaction, as well as the one of the reaction DEE + OH, are shown in figure 7.7.

In the system DME + OH, both the ip- and op-channel are indirect, complex-forming, channels according to reactions R11, R12 and R13 (cf. section 5.1). No relevant direct channels were found for this system. As a result, the abstractions of $H_{p,op}$ and $H_{p,ip}$ in the system DMM + OH are also expected to proceed via a prereactive complex, in which the OH radical is linked to one of the O atoms of the ether by a hydrogen bond. Moreover, the op-channel is expected to be energetically favored over the ip-channel.

The negative temperature dependence of k_4 between 230 and 370 K seems to be proven by the experimental data (cf. figure 7.6). Consequently, at least one complex-forming channel must exist in which the barrier of the forward reaction of the prereactive complex lies considerably lower than the reactants. Two reasons give rise to the assumption that this is the abstraction of H_s: First, no reasonable argument can be found for a significant change in the potential energies for the p,op- and p,ip-channel in comparison to the system DME + OH. Here, an energy barrier below the reactants was found for the op-channel in the calculations. However, the probable underestimation of this barrier by the theory has already been discussed by Carr *et al.* [96] and can be transferred to the calculations from ref. [157] as well. In contrast, the positive temperature dependence of k_2 is experimentally verified between 195 and 1470 K. This argues that the transition state energies of these channels lie

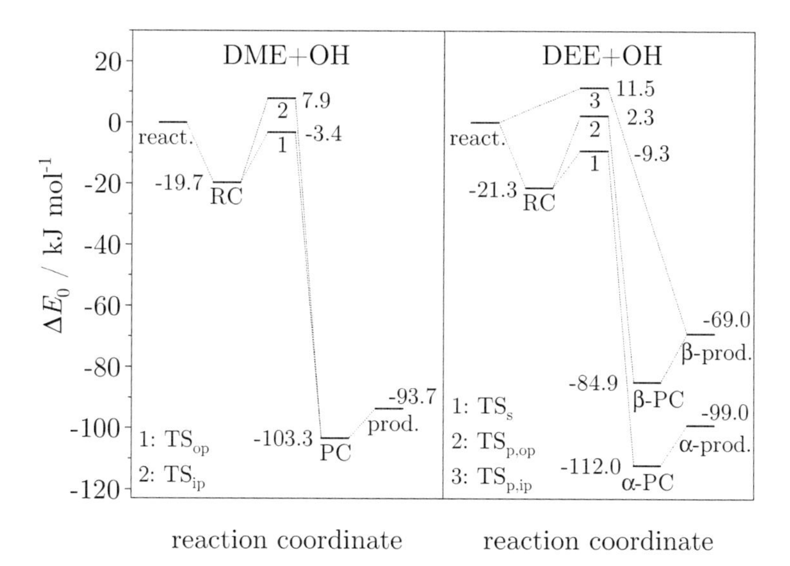

Figure 7.7: Potential energy diagrams (including zero-point energies) of the reactions DME + OH and DEE + OH calculated at CCSD(T)/cc-pV(T+Q)Z//CCSD/cc-pVDZ level of theory by Szöri and Kiecherer [18, 157, 182].

above the energies of the reactants. Second, it can be assumed that the bond dissociation energy (BDE) of the C–H-bond in the secondary position is lower in comparison to the primary positions. For the BDE of $C-H_s$ a value of 388.7 kJ mol^{-1} [203] is reported. The BDE of $C-H_p$ can be estimated with the help of the BDE of the C–H-bonds in DME, amounting to 402.1 kJ mol^{-1} [203]. This, and the better stabilization of the radical in the secondary position, leads to the conclusion that the transition state of the H_s abstraction is energetically favored over the one of the $H_{p,op}$ abstraction.

As only one distinguishable complex with a hydrogen bond between the OH radical and the ether O atom can be formed from the symmetrical DMM molecule, all indirect reaction channels must proceed via the same complex. In analogy to DME + OH and DEE + OH, the forward abstraction step is assumed to be irreversible under combustion-relevant conditions. Thus, the energies of potential product complexes and the separated products are not relevant for this discussion. A schematic potential energy diagram illustrating the considerations made so far is shown in figure 7.8.

To evaluate the differences in the temperature dependencies of k_4 obtained from the studies in this work and ref. [114], the examination of the rate coefficients of the indirect channels of DME + OH and DEE + OH which result from the potential energy diagrams of figure 7.7 is reasonable. Figure 7.9 shows an Arrhenius plot of the calculated values of Kiecherer [18, 183] compared to the experimentally obtained k_4.

The curvature obtained in ref. [114] is well described by the theoretically predicted rate

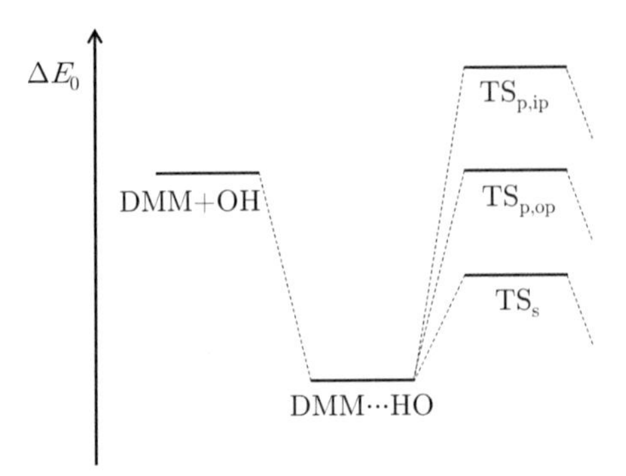

Figure 7.8: Schematic drawing of the derived characteristics of the potential energy diagram for reaction R4.

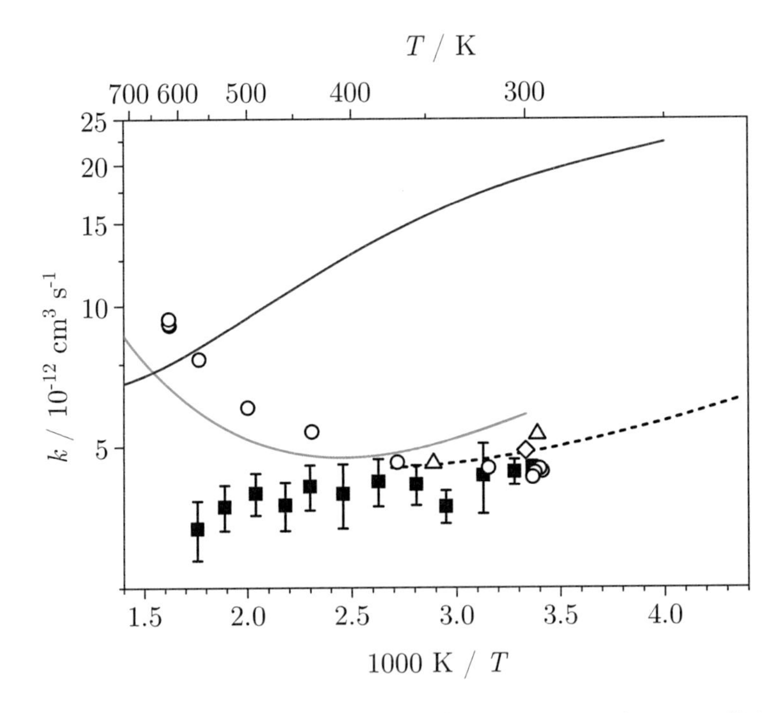

Figure 7.9: Arrhenius plot of k_4 (black), k_2 (green) and the sum of rate coefficients of the indirect channels of reaction R3 (blue): this work (■), Porter *et al.* [200] (- -), Wallington *et al.* [201] (△), Thüner *et al.* [202] (◇), Vovelle *et al.* [114] (○); Kiecherer [18] (—); Kiecherer [183] (—).

coefficient for the reaction DME + OH. Hence, if these experimental results give an adequate description, it can be assumed that the barrier of the H_s abstraction in DMM + OH is only slightly below the reactants. In the system DME + OH this energy barrier was calculated to be -3.5 kJ mol^{-1} [18, 157].

The temperature dependence for k_4 obtained in this work fits the one of the overall rate coefficient of the indirect DEE + OH channels well, although its absolute values are higher. This absolute position is influenced by many factors like, for example, well depth of the complex formation or the sum of states in the particular molecules. Thus, an exact agreement with the system DMM + OH cannot be expected. Nevertheless, the temperature dependence is mainly determined by the barrier height of the forward reaction relative to the reactants. Consequently, in the case of the adequacy of the experimental findings in this work, a barrier height similar to the one of the s-channel of DEE + OH should be valid. Here, a value of -9.3 kJ mol^{-1} was obtained in the calculations of Kiecherer [183].

As a result, both predicted temperature dependencies can be explained by the assumed potential energy diagram. Nevertheless, it can be deduced that the resulting Arrhenius behavior is highly dependent on the potential energy of the transition state TS_s. Thus, the knowledge of the precise barrier height of this reaction channel could help to validate one of the obtained Arrhenius profiles.

The discussion so far also implicates consequences for the pressure dependence of k_4 and its isotope effect. In the case of a higher barrier of the s-channel, similar to the one of the ip-channel of DME + OH from refs. [18, 157], a negligible pressure dependence and an isotope effect in the range of 3 to 4 is expected (cf. chapter 5). In the case of a lower barrier, like in the s-channel of the reaction DEE + OH [18, 182], a slight pressure dependence at pressures around or above 10 bar and an isotope effect of approx. 2 would be plausible (cf. chapter 6).

7.5 Conclusion and Outlook

The rate coefficient k_4 of the reaction DMM + OH was studied under autoignition-relevant conditions at temperatures between 297 and 570 K and 2, 5, and 10 bar pressure. A weak pressure dependence was observed. However, it lies in the range of the experimental error and could not be verified by a comparison with the experimental findings from other publications [114, 200–202]. Thus, an absence of a pressure dependence under the present conditions was concluded.

At moderate temperatures up to 370 K a good agreement of the rate coefficients from the literature [114, 200–202] and the results of this work was observed. Thus, it can be concluded that the kinetics of reaction R4 is well parametrized under these conditions. At temperatures higher than 400 K, the values of Vovelle *et al.* [114] and this work diverge significantly.

This differing behavior can influence the modeling of combustion processes. Especially the autoignition exhibits a high sensitivity on this reaction. Hence, more experimental studies over a broad temperature for a verification of one of the observed temperature dependencies are desirable. Depending on this validation, an adjustment of the applied parameters of k_4 in the model mechanisms from refs. [189, 191, 192] is recommendable.

In analogy to the theoretical findings regarding the reactions DME + OH [18, 157] and DEE + OH [18, 182, 183], general characteristics of the potential energy surface of reaction R4 could be deduced. It is assumed that the abstractions of all types of H atoms proceed via a prereactive complex, while the abstraction of H_s forms the main channel. The potential energy of its transition state TS_s is expected to be lower than the one of the reactants.

From these considerations, conclusions on the expected temperature dependence of k_4 can be derived. A potential energy of TS_s slightly below the reactants (≈-3 kJ mol^{-1}) is expected for an Arrhenius behavior according to ref. [114], while for the temperature dependence obtained in this work a significantly reduced energy can be assumed (≈-9 kJ mol^{-1}). Consequently, high level quantum chemical calculations of the potential energy surface of reaction R4 could contribute to a validation of the experimental observations. According to the findings in refs. [18, 157] and in chapter 5, it is advised against the use of the B3LYP functional for this purpose. Moreover, due to the high sensitivity of the characteristics of k_4 on the barrier heights, the application of costly coupled cluster methods with a large basis set is recommendable. In addition, indirect conclusions on the barrier height of the s-channel could be deduced from the study of a potential pressure dependence and the isotope effect of k_4 analogous to the studies of DME(-d6) + OH and DEE(-d10) + OH.

A Appendices

A.1 Appendix to Chapter 4

Table A.1: Overview of the experimental conditions in the measurements on the system HNO_3 + OH and the obtained pseudo-first order rate coefficients k_1'. The pressure inside the PLP/LIF cell is designated by P_{cell}, while $P_{bubbler}$ represents the pressure inside the bubbler.

T / K	P_{cell} / bar	$P_{bubbler}$ / bar	$[HNO_3]$ / cm^{-3}	k_1'/ s^{-1}
295	10	11.1	$1.33 \cdot 10^{17}$	22696
295	10	12.3	$1.22 \cdot 10^{17}$	19714
295	10	13.1	$1.15 \cdot 10^{17}$	19464
296	10	13.6	$1.12 \cdot 10^{17}$	15955
296	10	14.7	$1.04 \cdot 10^{17}$	16341
296	10	15.1	$1.00 \cdot 10^{17}$	15238
295	10	16.0	$9.54 \cdot 10^{16}$	15169
295	10	17.3	$8.84 \cdot 10^{16}$	12031
295	10	18.2	$8.40 \cdot 10^{16}$	15301
296	10	19.8	$7.94 \cdot 10^{16}$	11729
297	10	22.1	$7.15 \cdot 10^{16}$	11639
296	10	24.0	$6.61 \cdot 10^{16}$	11158
296	10	26.8	$6.02 \cdot 10^{16}$	10080
296	10	29.7	$5.52 \cdot 10^{16}$	7993
296	10	34.5	$4.90 \cdot 10^{16}$	7775
295	10	39.0	$4.44 \cdot 10^{16}$	7194
296	10	48.8	$3.63 \cdot 10^{16}$	6167
296	10	56.1	$3.36 \cdot 10^{16}$	6104
296	10	80.8	$2.59 \cdot 10^{16}$	3700

A.2 Appendix to Chapter 5

Table A.2: Overview of the experimental conditions in the measurements on the system DME-d6 + OH and the obtained pseudo-first order rate coefficients k'_{2d}. The horizontal lines indicate the differentiation between the individual series of measurements.

T / K	P / bar	[DME-d6] / cm^{-3}	[HNO_3] / cm^{-3}	k'_{2d} / s^{-1}
528	20.4	$1.35{\cdot}10^{17}$	$9.17{\cdot}10^{15}$	233379
527	20.2	$1.34{\cdot}10^{17}$	$9.12{\cdot}10^{15}$	210083
528	20.1	$1.33{\cdot}10^{17}$	$9.04{\cdot}10^{15}$	203902
527	20.0	$1.33{\cdot}10^{17}$	$9.03{\cdot}10^{15}$	211947
527	19.7	$1.31{\cdot}10^{17}$	$8.90{\cdot}10^{15}$	199542
528	19.6	$1.30{\cdot}10^{17}$	$8.81{\cdot}10^{15}$	215400
528	19.4	$1.29{\cdot}10^{17}$	$8.76{\cdot}10^{15}$	214862
527	19.3	$1.28{\cdot}10^{17}$	$8.69{\cdot}10^{15}$	212359
527	19.1	$1.27{\cdot}10^{17}$	$8.63{\cdot}10^{15}$	207755
527	18.9	$1.26{\cdot}10^{17}$	$8.56{\cdot}10^{15}$	191417
527	18.9	$1.25{\cdot}10^{17}$	$8.51{\cdot}10^{15}$	202167
527	18.7	$1.24{\cdot}10^{17}$	$8.44{\cdot}10^{15}$	196789
527	18.5	$1.23{\cdot}10^{17}$	$8.35{\cdot}10^{15}$	206301
527	18.3	$1.22{\cdot}10^{17}$	$8.26{\cdot}10^{15}$	221152
528	18.1	$1.20{\cdot}10^{17}$	$8.18{\cdot}10^{15}$	200807
527	17.8	$1.18{\cdot}10^{17}$	$8.04{\cdot}10^{15}$	183301
527	17.8	$1.18{\cdot}10^{17}$	$8.01{\cdot}10^{15}$	172391
527	17.6	$1.17{\cdot}10^{17}$	$7.92{\cdot}10^{15}$	184643
527	17.4	$1.15{\cdot}10^{17}$	$7.83{\cdot}10^{15}$	180518
527	17.3	$1.15{\cdot}10^{17}$	$7.79{\cdot}10^{15}$	191310
417	16.4	$1.38{\cdot}10^{17}$	$9.35{\cdot}10^{15}$	134561
417	16.3	$1.37{\cdot}10^{17}$	$9.27{\cdot}10^{15}$	147901
417	16.1	$1.36{\cdot}10^{17}$	$9.21{\cdot}10^{15}$	135727
417	16.1	$1.35{\cdot}10^{17}$	$9.16{\cdot}10^{15}$	136231
417	15.9	$1.33{\cdot}10^{17}$	$9.04{\cdot}10^{15}$	136472
416	15.8	$1.33{\cdot}10^{17}$	$9.01{\cdot}10^{15}$	139915
417	15.6	$1.32{\cdot}10^{17}$	$8.93{\cdot}10^{15}$	130443
417	15.5	$1.30{\cdot}10^{17}$	$8.84{\cdot}10^{15}$	121894
416	15.3	$1.29{\cdot}10^{17}$	$8.76{\cdot}10^{15}$	140088

T / K	P / bar	[DME-d6] / cm^{-3}	[HNO_3] / cm^{-3}	k'_{2d} / s^{-1}
417	15.2	$1.28 \cdot 10^{17}$	$8.68 \cdot 10^{15}$	123104
417	15.1	$1.27 \cdot 10^{17}$	$8.59 \cdot 10^{15}$	131832
416	14.9	$1.26 \cdot 10^{17}$	$8.56 \cdot 10^{15}$	128101
417	14.8	$1.24 \cdot 10^{17}$	$8.42 \cdot 10^{15}$	126125
417	14.7	$1.24 \cdot 10^{17}$	$8.39 \cdot 10^{15}$	130647
417	14.6	$1.22 \cdot 10^{17}$	$8.30 \cdot 10^{15}$	130083
416	14.4	$1.21 \cdot 10^{17}$	$8.22 \cdot 10^{15}$	123121
417	14.3	$1.20 \cdot 10^{17}$	$8.14 \cdot 10^{15}$	118408
417	14.1	$1.19 \cdot 10^{17}$	$8.08 \cdot 10^{15}$	127317
417	14.1	$1.18 \cdot 10^{17}$	$8.02 \cdot 10^{15}$	111108
417	13.9	$1.17 \cdot 10^{17}$	$7.94 \cdot 10^{15}$	118331
553	17.8	$1.28 \cdot 10^{17}$	$1.11 \cdot 10^{16}$	216038
554	17.6	$1.27 \cdot 10^{17}$	$1.10 \cdot 10^{16}$	242744
554	17.6	$1.28 \cdot 10^{17}$	$1.11 \cdot 10^{16}$	234382
554	17.4	$1.26 \cdot 10^{17}$	$1.09 \cdot 10^{16}$	231798
554	17.3	$1.25 \cdot 10^{17}$	$1.09 \cdot 10^{16}$	237706
554	17.1	$1.25 \cdot 10^{17}$	$1.08 \cdot 10^{16}$	226144
554	17.1	$1.24 \cdot 10^{17}$	$1.08 \cdot 10^{16}$	223631
553	16.8	$1.22 \cdot 10^{17}$	$1.06 \cdot 10^{16}$	214044
554	16.8	$1.22 \cdot 10^{17}$	$1.06 \cdot 10^{16}$	224022
554	16.8	$1.22 \cdot 10^{17}$	$1.06 \cdot 10^{16}$	226702
554	16.6	$1.21 \cdot 10^{17}$	$1.05 \cdot 10^{16}$	221873
554	16.5	$1.20 \cdot 10^{17}$	$1.04 \cdot 10^{16}$	219079
554	16.4	$1.19 \cdot 10^{17}$	$1.03 \cdot 10^{16}$	233181
553	16.2	$1.18 \cdot 10^{17}$	$1.02 \cdot 10^{16}$	209271
554	16.1	$1.17 \cdot 10^{17}$	$1.02 \cdot 10^{16}$	212335
554	15.9	$1.16 \cdot 10^{17}$	$1.01 \cdot 10^{16}$	211324
554	15.9	$1.15 \cdot 10^{17}$	$1.00 \cdot 10^{16}$	220206
554	15.8	$1.15 \cdot 10^{17}$	$9.94 \cdot 10^{15}$	229535
554	15.6	$1.14 \cdot 10^{17}$	$9.87 \cdot 10^{15}$	207266
554	15.6	$1.13 \cdot 10^{17}$	$9.81 \cdot 10^{15}$	204908
497	15.1	$1.23 \cdot 10^{17}$	$1.07 \cdot 10^{16}$	183765
497	15.0	$1.22 \cdot 10^{17}$	$1.05 \cdot 10^{16}$	194668
497	14.9	$1.21 \cdot 10^{17}$	$1.05 \cdot 10^{16}$	176331
497	14.8	$1.19 \cdot 10^{17}$	$1.04 \cdot 10^{16}$	171238

T / K	P / bar	[DME-d6] / cm^{-3}	[HNO_3] / cm^{-3}	k'_{2d} / s^{-1}
497	14.6	$1.19 \cdot 10^{17}$	$1.03 \cdot 10^{16}$	165412
499	14.6	$1.18 \cdot 10^{17}$	$1.02 \cdot 10^{16}$	179236
499	14.4	$1.16 \cdot 10^{17}$	$1.01 \cdot 10^{16}$	169403
499	14.3	$1.16 \cdot 10^{17}$	$1.00 \cdot 10^{16}$	170663
498	14.1	$1.15 \cdot 10^{17}$	$9.94 \cdot 10^{15}$	174063
497	14.1	$1.14 \cdot 10^{17}$	$9.91 \cdot 10^{15}$	171073
499	13.9	$1.13 \cdot 10^{17}$	$9.78 \cdot 10^{15}$	158031
499	13.9	$1.12 \cdot 10^{17}$	$9.71 \cdot 10^{15}$	173000
499	13.8	$1.11 \cdot 10^{17}$	$9.64 \cdot 10^{15}$	161418
497	13.6	$1.11 \cdot 10^{17}$	$9.59 \cdot 10^{15}$	159425
499	13.6	$1.09 \cdot 10^{17}$	$9.50 \cdot 10^{15}$	171567
499	13.4	$1.09 \cdot 10^{17}$	$9.43 \cdot 10^{15}$	170318
497	13.3	$1.07 \cdot 10^{17}$	$9.31 \cdot 10^{15}$	156147
499	13.2	$1.07 \cdot 10^{17}$	$9.25 \cdot 10^{15}$	170755
497	13.1	$1.06 \cdot 10^{17}$	$9.17 \cdot 10^{15}$	170236
497	12.9	$1.05 \cdot 10^{17}$	$9.10 \cdot 10^{15}$	160085
434	19.9	$1.74 \cdot 10^{17}$	$1.05 \cdot 10^{16}$	205070
434	19.8	$1.72 \cdot 10^{17}$	$1.04 \cdot 10^{16}$	224050
434	19.6	$1.71 \cdot 10^{17}$	$1.03 \cdot 10^{16}$	192237
434	19.6	$1.71 \cdot 10^{17}$	$1.03 \cdot 10^{16}$	200327
434	19.5	$1.70 \cdot 10^{17}$	$1.02 \cdot 10^{16}$	205106
434	19.4	$1.69 \cdot 10^{17}$	$1.02 \cdot 10^{16}$	194055
434	19.3	$1.68 \cdot 10^{17}$	$1.01 \cdot 10^{16}$	194358
434	19.1	$1.67 \cdot 10^{17}$	$1.00 \cdot 10^{16}$	191510
434	19.0	$1.65 \cdot 10^{17}$	$9.95 \cdot 10^{15}$	205441
434	18.9	$1.64 \cdot 10^{17}$	$9.87 \cdot 10^{15}$	158152
434	18.8	$1.63 \cdot 10^{17}$	$9.81 \cdot 10^{15}$	180518
434	18.6	$1.62 \cdot 10^{17}$	$9.76 \cdot 10^{15}$	189845
434	18.6	$1.61 \cdot 10^{17}$	$9.71 \cdot 10^{15}$	170619
434	18.4	$1.60 \cdot 10^{17}$	$9.60 \cdot 10^{15}$	181300
434	18.3	$1.59 \cdot 10^{17}$	$9.55 \cdot 10^{15}$	171431
434	18.1	$1.58 \cdot 10^{17}$	$9.50 \cdot 10^{15}$	171625
434	18.1	$1.57 \cdot 10^{17}$	$9.47 \cdot 10^{15}$	173097
435	17.9	$1.56 \cdot 10^{17}$	$9.36 \cdot 10^{15}$	174648
435	17.8	$1.55 \cdot 10^{17}$	$9.31 \cdot 10^{15}$	182197

T / K	P / bar	[DME-d6] / cm^{-3}	[HNO_3] / cm^{-3}	k'_{2d} / s^{-1}
435	17.6	$1.53{\cdot}10^{17}$	$9.23{\cdot}10^{15}$	175695
388	17.4	$1.69{\cdot}10^{17}$	$1.02{\cdot}10^{16}$	161575
388	17.2	$1.68{\cdot}10^{17}$	$1.01{\cdot}10^{16}$	171015
388	17.1	$1.67{\cdot}10^{17}$	$1.00{\cdot}10^{16}$	175525
388	17.0	$1.66{\cdot}10^{17}$	$9.96{\cdot}10^{15}$	149248
388	16.8	$1.64{\cdot}10^{17}$	$9.85{\cdot}10^{15}$	163460
388	16.7	$1.63{\cdot}10^{17}$	$9.79{\cdot}10^{15}$	151284
388	16.6	$1.61{\cdot}10^{17}$	$9.70{\cdot}10^{15}$	146280
388	16.4	$1.60{\cdot}10^{17}$	$9.64{\cdot}10^{15}$	152534
388	16.4	$1.59{\cdot}10^{17}$	$9.58{\cdot}10^{15}$	156786
388	16.3	$1.58{\cdot}10^{17}$	$9.53{\cdot}10^{15}$	147709
388	16.1	$1.57{\cdot}10^{17}$	$9.47{\cdot}10^{15}$	152748
388	16.1	$1.56{\cdot}10^{17}$	$9.41{\cdot}10^{15}$	140975
388	15.9	$1.55{\cdot}10^{17}$	$9.35{\cdot}10^{15}$	154926
388	15.8	$1.54{\cdot}10^{17}$	$9.26{\cdot}10^{15}$	145151
388	15.7	$1.53{\cdot}10^{17}$	$9.21{\cdot}10^{15}$	140784
388	15.6	$1.52{\cdot}10^{17}$	$9.12{\cdot}10^{15}$	146264
387	15.4	$1.50{\cdot}10^{17}$	$9.03{\cdot}10^{15}$	148290
388	15.3	$1.49{\cdot}10^{17}$	$8.97{\cdot}10^{15}$	133368
388	15.1	$1.48{\cdot}10^{17}$	$8.88{\cdot}10^{15}$	139431
388	15.1	$1.47{\cdot}10^{17}$	$8.85{\cdot}10^{15}$	133425

A.3 Appendix to Chapter 6

Table A.3: Overview of the experimental conditions in the measurements on the system DEE + OH and the obtained pseudo-first order rate coefficients k_3'. The horizontal lines indicate the differentiation between the individual series of measurements.

T / K	P / bar	[DEE] / cm^{-3}	[HNO_3] / cm^{-3}	k_3' / s^{-1}
295	2	$1.06 \cdot 10^{16}$	$3.61 \cdot 10^{15}$	146912
295	2	$9.63 \cdot 10^{15}$	$3.61 \cdot 10^{15}$	121729
295	2	$8.57 \cdot 10^{15}$	$3.61 \cdot 10^{15}$	116346
295	2	$7.56 \cdot 10^{15}$	$3.61 \cdot 10^{15}$	99155
295	2	$6.50 \cdot 10^{15}$	$3.61 \cdot 10^{15}$	83643
295	2	$5.49 \cdot 10^{15}$	$3.61 \cdot 10^{15}$	72324
295	2	$4.43 \cdot 10^{15}$	$3.61 \cdot 10^{15}$	60603
295	2	$3.42 \cdot 10^{15}$	$3.61 \cdot 10^{15}$	43998
295	2	$2.36 \cdot 10^{15}$	$3.61 \cdot 10^{15}$	35390
295	2	$1.18 \cdot 10^{15}$	$3.61 \cdot 10^{15}$	17941
295	2	0	$3.61 \cdot 10^{15}$	7208
454	2	0	$6.52 \cdot 10^{15}$	3672
454	2	$8.54 \cdot 10^{14}$	$6.52 \cdot 10^{15}$	13870
454	2	$1.58 \cdot 10^{15}$	$6.52 \cdot 10^{15}$	20497
454	2	$2.35 \cdot 10^{15}$	$6.52 \cdot 10^{15}$	23643
454	2	$3.07 \cdot 10^{15}$	$6.52 \cdot 10^{15}$	41981
454	2	$3.84 \cdot 10^{15}$	$6.52 \cdot 10^{15}$	39414
454	2	$4.57 \cdot 10^{15}$	$6.52 \cdot 10^{15}$	58908
454	2	$5.34 \cdot 10^{15}$	$6.52 \cdot 10^{15}$	63791
454	2	$6.06 \cdot 10^{15}$	$6.51 \cdot 10^{15}$	73890
454	2	$6.83 \cdot 10^{15}$	$6.52 \cdot 10^{15}$	103566
454	2	$7.68 \cdot 10^{15}$	$6.52 \cdot 10^{15}$	87688
454	10	0	$3.26 \cdot 10^{16}$	28284
454	10	$4.27 \cdot 10^{15}$	$3.26 \cdot 10^{16}$	103280
454	10	$7.90 \cdot 10^{15}$	$3.26 \cdot 10^{16}$	135293
454	10	$1.17 \cdot 10^{16}$	$3.26 \cdot 10^{16}$	170299
454	10	$1.54 \cdot 10^{16}$	$3.26 \cdot 10^{16}$	280214
454	10	$1.92 \cdot 10^{16}$	$3.26 \cdot 10^{16}$	244617
454	10	$2.28 \cdot 10^{16}$	$3.26 \cdot 10^{16}$	330641
454	10	$2.67 \cdot 10^{16}$	$3.26 \cdot 10^{16}$	304947

T / K	P / bar	[DEE] / cm^{-3}	[HNO_3] / cm^{-3}	k_3' / s^{-1}
454	10	$3.03 \cdot 10^{16}$	$3.26 \cdot 10^{16}$	414379
454	10	$3.42 \cdot 10^{16}$	$3.26 \cdot 10^{16}$	347552
454	10	$3.84 \cdot 10^{16}$	$3.26 \cdot 10^{16}$	492554
422	10	$4.13 \cdot 10^{16}$	$3.51 \cdot 10^{16}$	487802
422	10	$3.74 \cdot 10^{16}$	$3.50 \cdot 10^{16}$	497299
422	10	$3.33 \cdot 10^{16}$	$3.50 \cdot 10^{16}$	396798
423	10	$2.93 \cdot 10^{16}$	$3.50 \cdot 10^{16}$	330739
423	10	$2.52 \cdot 10^{16}$	$3.50 \cdot 10^{16}$	374731
423	10	$2.13 \cdot 10^{16}$	$3.49 \cdot 10^{16}$	282788
424	10	$1.72 \cdot 10^{16}$	$3.49 \cdot 10^{16}$	227931
424	10	$1.33 \cdot 10^{16}$	$3.49 \cdot 10^{16}$	159945
424	10	$9.14 \cdot 10^{15}$	$3.49 \cdot 10^{16}$	143264
424	10	$4.57 \cdot 10^{15}$	$3.49 \cdot 10^{16}$	71877
425	10	0	$3.48 \cdot 10^{16}$	21957
427	2	$8.17 \cdot 10^{15}$	$6.93 \cdot 10^{15}$	141331
427	2	$7.39 \cdot 10^{15}$	$6.92 \cdot 10^{15}$	96921
428	2	$6.57 \cdot 10^{15}$	$6.92 \cdot 10^{15}$	78601
428	2	$5.80 \cdot 10^{15}$	$6.92 \cdot 10^{15}$	67239
428	2	$4.98 \cdot 10^{15}$	$6.92 \cdot 10^{15}$	62417
428	2	$4.21 \cdot 10^{15}$	$6.91 \cdot 10^{15}$	48936
428	2	$3.40 \cdot 10^{15}$	$6.91 \cdot 10^{15}$	39612
428	2	$2.63 \cdot 10^{15}$	$6.91 \cdot 10^{15}$	33918
428	2	$1.81 \cdot 10^{15}$	$6.91 \cdot 10^{15}$	19452
428	2	$9.05 \cdot 10^{14}$	$6.91 \cdot 10^{15}$	13368
428	2	0	$6.91 \cdot 10^{15}$	3319
295	5	0	$2.51 \cdot 10^{16}$	24386
295	5	$3.29 \cdot 10^{15}$	$2.51 \cdot 10^{16}$	66268
295	5	$6.08 \cdot 10^{15}$	$2.51 \cdot 10^{16}$	91344
295	5	$9.03 \cdot 10^{15}$	$2.51 \cdot 10^{16}$	121783
295	5	$1.18 \cdot 10^{16}$	$2.51 \cdot 10^{16}$	164986
295	5	$1.48 \cdot 10^{16}$	$2.51 \cdot 10^{16}$	168645
295	5	$1.76 \cdot 10^{16}$	$2.51 \cdot 10^{16}$	222370
295	5	$2.05 \cdot 10^{16}$	$2.51 \cdot 10^{16}$	306375
295	5	$2.33 \cdot 10^{16}$	$2.51 \cdot 10^{16}$	346340
295	5	$2.63 \cdot 10^{16}$	$2.50 \cdot 10^{16}$	288274

T / K	P / bar	[DEE] / cm^{-3}	[HNO_3] / cm^{-3}	k'_3 / s^{-1}
295	5	$2.95 \cdot 10^{16}$	$2.50 \cdot 10^{16}$	355360
304	2	0	$9.72 \cdot 10^{15}$	6717
304	2	$1.27 \cdot 10^{15}$	$9.72 \cdot 10^{15}$	21823
304	2	$2.36 \cdot 10^{15}$	$9.72 \cdot 10^{15}$	41918
305	2	$3.50 \cdot 10^{15}$	$9.71 \cdot 10^{15}$	50721
305	2	$4.58 \cdot 10^{15}$	$9.71 \cdot 10^{15}$	61286
305	2	$5.73 \cdot 10^{15}$	$9.71 \cdot 10^{15}$	85112
305	2	$6.81 \cdot 10^{15}$	$9.71 \cdot 10^{15}$	87418
305	2	$7.95 \cdot 10^{15}$	$9.71 \cdot 10^{15}$	93463
305	2	$9.03 \cdot 10^{15}$	$9.71 \cdot 10^{15}$	110075
305	2	$1.02 \cdot 10^{16}$	$9.71 \cdot 10^{15}$	158725
305	2	$1.14 \cdot 10^{16}$	$9.71 \cdot 10^{15}$	176402
317	2	0	$8.87 \cdot 10^{15}$	3980
317	2	$7.51 \cdot 10^{14}$	$8.87 \cdot 10^{15}$	11562
317	2	$1.39 \cdot 10^{15}$	$8.86 \cdot 10^{15}$	20389
318	2	$2.06 \cdot 10^{15}$	$8.85 \cdot 10^{15}$	27626
318	2	$2.70 \cdot 10^{15}$	$8.85 \cdot 10^{15}$	36852
318	2	$3.37 \cdot 10^{15}$	$8.85 \cdot 10^{15}$	43333
318	2	$4.01 \cdot 10^{15}$	$8.84 \cdot 10^{15}$	51015
318	2	$4.68 \cdot 10^{15}$	$8.84 \cdot 10^{15}$	65839
318	2	$5.32 \cdot 10^{15}$	$8.84 \cdot 10^{15}$	65281
318	2	$5.99 \cdot 10^{15}$	$8.83 \cdot 10^{15}$	75220
318	2	$6.73 \cdot 10^{15}$	$8.83 \cdot 10^{15}$	92364
318	5	$1.68 \cdot 10^{16}$	$2.21 \cdot 10^{16}$	219272
319	5	$1.52 \cdot 10^{16}$	$2.21 \cdot 10^{16}$	202547
319	5	$1.35 \cdot 10^{16}$	$2.21 \cdot 10^{16}$	190608
319	5	$1.20 \cdot 10^{16}$	$2.21 \cdot 10^{16}$	166656
319	5	$1.03 \cdot 10^{16}$	$2.20 \cdot 10^{16}$	143057
319	5	$8.68 \cdot 10^{15}$	$2.20 \cdot 10^{16}$	112202
319	5	$7.00 \cdot 10^{15}$	$2.20 \cdot 10^{16}$	91156
319	5	$5.41 \cdot 10^{15}$	$2.20 \cdot 10^{16}$	67966
319	5	$3.73 \cdot 10^{15}$	$2.20 \cdot 10^{16}$	51718
319	5	$1.87 \cdot 10^{15}$	$2.20 \cdot 10^{16}$	30910
319	5	0	$2.20 \cdot 10^{16}$	10970
319	10	0	$4.40 \cdot 10^{16}$	28048

T / K	P / bar	[DEE] / cm^{-3}	[HNO_3] / cm^{-3}	k'_3 / s^{-1}
319	10	$3.73 \cdot 10^{15}$	$4.40 \cdot 10^{16}$	68493
319	10	$6.90 \cdot 10^{15}$	$4.40 \cdot 10^{16}$	112318
319	10	$1.03 \cdot 10^{16}$	$4.40 \cdot 10^{16}$	179774
319	10	$1.34 \cdot 10^{16}$	$4.40 \cdot 10^{16}$	197644
319	10	$1.68 \cdot 10^{16}$	$4.40 \cdot 10^{16}$	236723
319	10	$1.99 \cdot 10^{16}$	$4.40 \cdot 10^{16}$	278178
319	10	$2.33 \cdot 10^{16}$	$4.40 \cdot 10^{16}$	319805
320	10	$2.65 \cdot 10^{16}$	$4.40 \cdot 10^{16}$	345976
320	10	$2.98 \cdot 10^{16}$	$4.40 \cdot 10^{16}$	399679
320	10	$3.35 \cdot 10^{16}$	$4.40 \cdot 10^{16}$	524949
334	10	$3.21 \cdot 10^{16}$	$4.21 \cdot 10^{16}$	492389
334	10	$2.91 \cdot 10^{16}$	$4.21 \cdot 10^{16}$	370938
334	10	$2.58 \cdot 10^{16}$	$4.20 \cdot 10^{16}$	364821
334	10	$2.28 \cdot 10^{16}$	$4.20 \cdot 10^{16}$	303948
335	10	$1.96 \cdot 10^{16}$	$4.20 \cdot 10^{16}$	256941
335	10	$1.65 \cdot 10^{16}$	$4.20 \cdot 10^{16}$	230681
335	10	$1.33 \cdot 10^{16}$	$4.19 \cdot 10^{16}$	193984
335	10	$1.03 \cdot 10^{16}$	$4.19 \cdot 10^{16}$	161057
335	10	$7.10 \cdot 10^{15}$	$4.19 \cdot 10^{16}$	108351
336	10	$3.55 \cdot 10^{15}$	$4.19 \cdot 10^{16}$	66236
336	10	0	$4.19 \cdot 10^{16}$	30243
336	5	0	$2.09 \cdot 10^{16}$	10615
336	5	$1.77 \cdot 10^{15}$	$2.09 \cdot 10^{16}$	30870
336	5	$3.28 \cdot 10^{15}$	$2.09 \cdot 10^{16}$	53728
336	5	$4.87 \cdot 10^{15}$	$2.09 \cdot 10^{16}$	69320
336	5	$6.38 \cdot 10^{15}$	$2.09 \cdot 10^{16}$	89685
336	5	$7.97 \cdot 10^{15}$	$2.09 \cdot 10^{16}$	116500
336	5	$9.47 \cdot 10^{15}$	$2.09 \cdot 10^{16}$	126599
336	5	$1.11 \cdot 10^{16}$	$2.09 \cdot 10^{16}$	153411
336	5	$1.26 \cdot 10^{16}$	$2.09 \cdot 10^{16}$	173982
336	5	$1.42 \cdot 10^{16}$	$2.09 \cdot 10^{16}$	183145
336	5	$1.59 \cdot 10^{16}$	$2.09 \cdot 10^{16}$	214989
337	2	$6.37 \cdot 10^{15}$	$8.35 \cdot 10^{15}$	77288
337	2	$5.77 \cdot 10^{15}$	$8.35 \cdot 10^{15}$	66193
337	2	$5.13 \cdot 10^{15}$	$8.35 \cdot 10^{15}$	70820

T / K	P / bar	[DEE] / cm^{-3}	[HNO_3] / cm^{-3}	k'_3 / s^{-1}
337	2	$4.53 \cdot 10^{15}$	$8.35 \cdot 10^{15}$	62061
337	2	$3.89 \cdot 10^{15}$	$8.35 \cdot 10^{15}$	48088
337	2	$3.29 \cdot 10^{15}$	$8.35 \cdot 10^{15}$	41433
337	2	$2.65 \cdot 10^{15}$	$8.35 \cdot 10^{15}$	37696
337	2	$2.05 \cdot 10^{15}$	$8.35 \cdot 10^{15}$	25800
337	2	$1.41 \cdot 10^{15}$	$8.35 \cdot 10^{15}$	18881
337	2	$7.07 \cdot 10^{14}$	$8.35 \cdot 10^{15}$	11780
337	2	0	$8.35 \cdot 10^{15}$	3777
357	2	0	$7.87 \cdot 10^{15}$	3337
357	2	$6.66 \cdot 10^{14}$	$7.87 \cdot 10^{15}$	9518
357	2	$1.23 \cdot 10^{15}$	$7.87 \cdot 10^{15}$	15320
357	2	$1.83 \cdot 10^{15}$	$7.87 \cdot 10^{15}$	23666
357	2	$2.40 \cdot 10^{15}$	$7.87 \cdot 10^{15}$	30884
357	2	$3.00 \cdot 10^{15}$	$7.87 \cdot 10^{15}$	36194
357	2	$3.57 \cdot 10^{15}$	$7.87 \cdot 10^{15}$	44695
357	2	$4.17 \cdot 10^{15}$	$7.87 \cdot 10^{15}$	57509
357	2	$4.74 \cdot 10^{15}$	$7.87 \cdot 10^{15}$	61319
357	2	$5.34 \cdot 10^{15}$	$7.87 \cdot 10^{15}$	71255
357	2	$6.00 \cdot 10^{15}$	$7.88 \cdot 10^{15}$	77848
357	5	$1.50 \cdot 10^{16}$	$1.97 \cdot 10^{16}$	202640
357	5	$1.36 \cdot 10^{16}$	$1.97 \cdot 10^{16}$	161876
357	5	$1.21 \cdot 10^{16}$	$1.97 \cdot 10^{16}$	156970
357	5	$1.07 \cdot 10^{16}$	$1.97 \cdot 10^{16}$	131591
357	5	$9.18 \cdot 10^{15}$	$1.97 \cdot 10^{16}$	107379
357	5	$7.76 \cdot 10^{15}$	$1.97 \cdot 10^{16}$	94222
357	5	$6.26 \cdot 10^{15}$	$1.97 \cdot 10^{16}$	73750
357	5	$4.84 \cdot 10^{15}$	$1.97 \cdot 10^{16}$	56895
357	5	$3.34 \cdot 10^{15}$	$1.97 \cdot 10^{16}$	40753
356	5	$1.67 \cdot 10^{15}$	$1.97 \cdot 10^{16}$	25489
356	5	0	$1.97 \cdot 10^{16}$	7723
356	10	0	$3.94 \cdot 10^{16}$	26257
356	10	$3.34 \cdot 10^{15}$	$3.94 \cdot 10^{16}$	70581
356	10	$6.18 \cdot 10^{15}$	$3.94 \cdot 10^{16}$	96982
356	10	$9.19 \cdot 10^{15}$	$3.95 \cdot 10^{16}$	138596
356	10	$1.20 \cdot 10^{16}$	$3.95 \cdot 10^{16}$	162691

T / K	P / bar	[DEE] / cm^{-3}	[HNO_3] / cm^{-3}	k_3' / s^{-1}
356	10	$1.50 \cdot 10^{16}$	$3.95 \cdot 10^{16}$	226537
356	10	$1.79 \cdot 10^{16}$	$3.95 \cdot 10^{16}$	247158
356	10	$2.09 \cdot 10^{16}$	$3.95 \cdot 10^{16}$	296445
356	10	$2.37 \cdot 10^{16}$	$3.95 \cdot 10^{16}$	305714
356	10	$2.68 \cdot 10^{16}$	$3.95 \cdot 10^{16}$	388037
356	10	$3.01 \cdot 10^{16}$	$3.95 \cdot 10^{16}$	383703
378	5	$1.42 \cdot 10^{16}$	$1.86 \cdot 10^{16}$	177136
378	5	$1.28 \cdot 10^{16}$	$1.86 \cdot 10^{16}$	170180
378	5	$1.14 \cdot 10^{16}$	$1.86 \cdot 10^{16}$	157891
378	5	$1.01 \cdot 10^{16}$	$1.86 \cdot 10^{16}$	156540
378	5	$8.66 \cdot 10^{15}$	$1.86 \cdot 10^{16}$	132081
378	5	$7.32 \cdot 10^{15}$	$1.86 \cdot 10^{16}$	109496
378	5	$5.90 \cdot 10^{15}$	$1.86 \cdot 10^{16}$	86417
378	5	$4.57 \cdot 10^{15}$	$1.86 \cdot 10^{16}$	79614
378	5	$3.15 \cdot 10^{15}$	$1.86 \cdot 10^{16}$	65274
378	5	$1.58 \cdot 10^{15}$	$1.86 \cdot 10^{16}$	41346
378	5	0	$1.86 \cdot 10^{16}$	18156
378	2	0	$7.44 \cdot 10^{15}$	4150
378	2	$6.30 \cdot 10^{14}$	$7.44 \cdot 10^{15}$	11555
378	2	$1.17 \cdot 10^{15}$	$7.44 \cdot 10^{15}$	17617
378	2	$1.73 \cdot 10^{15}$	$7.44 \cdot 10^{15}$	25041
378	2	$2.27 \cdot 10^{15}$	$7.45 \cdot 10^{15}$	32442
378	2	$2.84 \cdot 10^{15}$	$7.45 \cdot 10^{15}$	39157
378	2	$3.38 \cdot 10^{15}$	$7.45 \cdot 10^{15}$	47434
377	2	$3.94 \cdot 10^{15}$	$7.45 \cdot 10^{15}$	59095
378	2	$4.48 \cdot 10^{15}$	$7.45 \cdot 10^{15}$	63338
377	2	$5.05 \cdot 10^{15}$	$7.45 \cdot 10^{15}$	61350
377	2	$5.68 \cdot 10^{15}$	$7.45 \cdot 10^{15}$	86375
400	5	$1.34 \cdot 10^{16}$	$1.76 \cdot 10^{16}$	219307
400	5	$1.21 \cdot 10^{16}$	$1.76 \cdot 10^{16}$	199357
400	5	$1.08 \cdot 10^{16}$	$1.76 \cdot 10^{16}$	171646
400	5	$9.52 \cdot 10^{15}$	$1.76 \cdot 10^{16}$	144011
400	5	$8.18 \cdot 10^{15}$	$1.76 \cdot 10^{16}$	133487
400	5	$6.91 \cdot 10^{15}$	$1.76 \cdot 10^{16}$	102731
400	5	$5.58 \cdot 10^{15}$	$1.76 \cdot 10^{16}$	100285

T / K	P / bar	[DEE] / cm^{-3}	[HNO_3] / cm^{-3}	k'_3 / s^{-1}
400	5	$4.31\cdot10^{15}$	$1.76\cdot10^{16}$	71396
400	5	$2.97\cdot10^{15}$	$1.75\cdot10^{16}$	54288
400	5	$1.49\cdot10^{15}$	$1.76\cdot10^{16}$	37013
400	5	0	$1.76\cdot10^{16}$	13544
400	2	0	$7.03\cdot10^{15}$	3907
400	2	$5.95\cdot10^{14}$	$7.03\cdot10^{15}$	12815
400	2	$1.10\cdot10^{15}$	$7.03\cdot10^{15}$	17093
400	2	$1.64\cdot10^{15}$	$7.03\cdot10^{15}$	22473
400	2	$2.14\cdot10^{15}$	$7.03\cdot10^{15}$	36205
400	2	$2.68\cdot10^{15}$	$7.03\cdot10^{15}$	34412
400	2	$3.19\cdot10^{15}$	$7.03\cdot10^{15}$	41296
400	2	$3.72\cdot10^{15}$	$7.03\cdot10^{15}$	42518
400	2	$4.23\cdot10^{15}$	$7.03\cdot10^{15}$	59543
400	2	$4.77\cdot10^{15}$	$7.04\cdot10^{15}$	55191
400	2	$5.37\cdot10^{15}$	$7.04\cdot10^{15}$	71776
488	5	$1.10\cdot10^{16}$	$1.44\cdot10^{16}$	151715
488	5	$9.94\cdot10^{15}$	$1.44\cdot10^{16}$	128851
488	5	$8.84\cdot10^{15}$	$1.44\cdot10^{16}$	132427
489	5	$7.80\cdot10^{15}$	$1.44\cdot10^{16}$	129216
489	5	$6.70\cdot10^{15}$	$1.44\cdot10^{16}$	117917
489	5	$5.66\cdot10^{15}$	$1.44\cdot10^{16}$	107122
489	5	$4.57\cdot10^{15}$	$1.44\cdot10^{16}$	86895
489	5	$3.53\cdot10^{15}$	$1.44\cdot10^{16}$	80278
489	5	$2.44\cdot10^{15}$	$1.44\cdot10^{16}$	72264
489	5	$1.22\cdot10^{15}$	$1.44\cdot10^{16}$	42471
489	5	0	$1.44\cdot10^{16}$	25979
294	10	0	$4.92\cdot10^{16}$	45446
294	10	$2.04\cdot10^{15}$	$4.92\cdot10^{16}$	71668
294	10	$3.78\cdot10^{15}$	$4.92\cdot10^{16}$	103950
294	10	$5.62\cdot10^{15}$	$4.92\cdot10^{16}$	128904
294	10	$7.35\cdot10^{15}$	$4.92\cdot10^{16}$	146839
294	10	$9.19\cdot10^{15}$	$4.92\cdot10^{16}$	174316
294	10	$1.09\cdot10^{16}$	$4.92\cdot10^{16}$	181967
294	10	$1.28\cdot10^{16}$	$4.92\cdot10^{16}$	192384
294	10	$1.45\cdot10^{16}$	$4.91\cdot10^{16}$	224139

T / K	P / bar	[DEE] / cm^{-3}	[HNO_3] / cm^{-3}	k'_3 / s^{-1}
294	10	$1.63 \cdot 10^{16}$	$4.91 \cdot 10^{16}$	283391
294	10	$1.84 \cdot 10^{16}$	$4.91 \cdot 10^{16}$	263906
304	10	$1.78 \cdot 10^{16}$	$4.75 \cdot 10^{16}$	225455
305	10	$1.61 \cdot 10^{16}$	$4.75 \cdot 10^{16}$	231230
305	10	$1.43 \cdot 10^{16}$	$4.75 \cdot 10^{16}$	207501
305	10	$1.26 \cdot 10^{16}$	$4.74 \cdot 10^{16}$	230756
305	10	$1.08 \cdot 10^{16}$	$4.74 \cdot 10^{16}$	197120
305	10	$9.16 \cdot 10^{15}$	$4.74 \cdot 10^{16}$	142276
305	10	$7.39 \cdot 10^{15}$	$4.74 \cdot 10^{16}$	130700
305	10	$5.71 \cdot 10^{15}$	$4.74 \cdot 10^{16}$	124248
305	10	$3.94 \cdot 10^{15}$	$4.74 \cdot 10^{16}$	90981
305	10	$1.97 \cdot 10^{15}$	$4.74 \cdot 10^{16}$	71092
305	10	0	$4.74 \cdot 10^{16}$	49936
377	10	0	$3.83 \cdot 10^{16}$	65487
378	10	$1.59 \cdot 10^{15}$	$3.83 \cdot 10^{16}$	105666
378	10	$2.94 \cdot 10^{15}$	$3.83 \cdot 10^{16}$	145131
378	10	$4.37 \cdot 10^{15}$	$3.82 \cdot 10^{16}$	169219
378	10	$5.72 \cdot 10^{15}$	$3.82 \cdot 10^{16}$	200989
378	10	$7.15 \cdot 10^{15}$	$3.82 \cdot 10^{16}$	207078
378	10	$8.50 \cdot 10^{15}$	$3.82 \cdot 10^{16}$	225014
378	10	$9.93 \cdot 10^{15}$	$3.83 \cdot 10^{16}$	246121
378	10	$1.13 \cdot 10^{16}$	$3.83 \cdot 10^{16}$	248683
378	10	$1.27 \cdot 10^{16}$	$3.83 \cdot 10^{16}$	293707
377	10	$1.43 \cdot 10^{16}$	$3.83 \cdot 10^{16}$	332597
305	5	0	$2.26 \cdot 10^{16}$	35175
305	5	$9.73 \cdot 10^{14}$	$2.26 \cdot 10^{16}$	50760
305	5	$1.80 \cdot 10^{15}$	$2.26 \cdot 10^{16}$	62028
306	5	$2.67 \cdot 10^{15}$	$2.25 \cdot 10^{16}$	76675
306	5	$3.50 \cdot 10^{15}$	$2.25 \cdot 10^{16}$	85125
305	5	$4.38 \cdot 10^{15}$	$2.26 \cdot 10^{16}$	100196
305	5	$5.21 \cdot 10^{15}$	$2.26 \cdot 10^{16}$	114549
305	5	$6.08 \cdot 10^{15}$	$2.26 \cdot 10^{16}$	129547
305	5	$6.90 \cdot 10^{15}$	$2.25 \cdot 10^{16}$	136321
305	5	$7.79 \cdot 10^{15}$	$2.26 \cdot 10^{16}$	152904
305	5	$8.77 \cdot 10^{15}$	$2.26 \cdot 10^{16}$	161026

T / K	P / bar	[DEE] / cm^{-3}	[HNO_3] / cm^{-3}	k'_3 / s^{-1}
400	10	$1.33 \cdot 10^{16}$	$3.44 \cdot 10^{16}$	215933
401	10	$1.21 \cdot 10^{16}$	$3.44 \cdot 10^{16}$	208625
401	10	$1.07 \cdot 10^{16}$	$3.43 \cdot 10^{16}$	188576
401	10	$9.47 \cdot 10^{15}$	$3.43 \cdot 10^{16}$	165864
401	10	$8.13 \cdot 10^{15}$	$3.43 \cdot 10^{16}$	146302
401	10	$6.89 \cdot 10^{15}$	$3.43 \cdot 10^{16}$	140671
400	10	$5.56 \cdot 10^{15}$	$3.44 \cdot 10^{16}$	109954
400	10	$4.30 \cdot 10^{15}$	$3.44 \cdot 10^{16}$	98842
400	10	$2.97 \cdot 10^{15}$	$3.44 \cdot 10^{16}$	78409
400	10	$1.48 \cdot 10^{15}$	$3.44 \cdot 10^{16}$	59367
400	10	0	$3.44 \cdot 10^{16}$	38643
426	5	0	$1.61 \cdot 10^{16}$	12152
426	5	$6.97 \cdot 10^{14}$	$1.62 \cdot 10^{16}$	18357
426	5	$1.29 \cdot 10^{15}$	$1.61 \cdot 10^{16}$	25851
426	5	$1.92 \cdot 10^{15}$	$1.62 \cdot 10^{16}$	35015
425	5	$2.51 \cdot 10^{15}$	$1.62 \cdot 10^{16}$	42875
425	5	$3.14 \cdot 10^{15}$	$1.62 \cdot 10^{16}$	50433
425	5	$3.74 \cdot 10^{15}$	$1.62 \cdot 10^{16}$	60821
425	5	$4.36 \cdot 10^{15}$	$1.62 \cdot 10^{16}$	65608
425	5	$4.96 \cdot 10^{15}$	$1.62 \cdot 10^{16}$	75810
425	5	$5.58 \cdot 10^{15}$	$1.62 \cdot 10^{16}$	80201
426	5	$6.27 \cdot 10^{15}$	$1.62 \cdot 10^{16}$	92010
454	5	$5.89 \cdot 10^{15}$	$1.52 \cdot 10^{16}$	85507
453	5	$5.34 \cdot 10^{15}$	$1.52 \cdot 10^{16}$	73619
453	5	$4.75 \cdot 10^{15}$	$1.52 \cdot 10^{16}$	68103
454	5	$4.19 \cdot 10^{15}$	$1.52 \cdot 10^{16}$	60665
454	5	$3.60 \cdot 10^{15}$	$1.52 \cdot 10^{16}$	52575
454	5	$3.04 \cdot 10^{15}$	$1.51 \cdot 10^{16}$	51280
454	5	$2.45 \cdot 10^{15}$	$1.51 \cdot 10^{16}$	38739
454	5	$1.90 \cdot 10^{15}$	$1.52 \cdot 10^{16}$	30951
454	5	$1.31 \cdot 10^{15}$	$1.52 \cdot 10^{16}$	24013
453	5	$6.55 \cdot 10^{14}$	$1.52 \cdot 10^{16}$	18313
453	5	0	$1.52 \cdot 10^{16}$	10991
488	2	0	$5.64 \cdot 10^{15}$	5123
489	2	$2.43 \cdot 10^{14}$	$5.63 \cdot 10^{15}$	8382

T / K	P / bar	[DEE] / cm^{-3}	[HNO_3] / cm^{-3}	k'_3 / s^{-1}
489	2	$4.49 \cdot 10^{14}$	$5.63 \cdot 10^{15}$	11284
489	2	$6.67 \cdot 10^{14}$	$5.63 \cdot 10^{15}$	14592
489	2	$8.74 \cdot 10^{14}$	$5.63 \cdot 10^{15}$	16399
489	2	$1.09 \cdot 10^{15}$	$5.63 \cdot 10^{15}$	18022
489	2	$1.30 \cdot 10^{15}$	$5.63 \cdot 10^{15}$	21358
488	2	$1.52 \cdot 10^{15}$	$5.64 \cdot 10^{15}$	23364
488	2	$1.73 \cdot 10^{15}$	$5.64 \cdot 10^{15}$	27173
489	2	$1.94 \cdot 10^{15}$	$5.63 \cdot 10^{15}$	28123
488	2	$2.19 \cdot 10^{15}$	$5.64 \cdot 10^{15}$	30309
530	2	$2.02 \cdot 10^{15}$	$5.19 \cdot 10^{15}$	29019
529	2	$1.83 \cdot 10^{15}$	$5.20 \cdot 10^{15}$	27855
529	2	$1.63 \cdot 10^{15}$	$5.20 \cdot 10^{15}$	24698
530	2	$1.43 \cdot 10^{15}$	$5.20 \cdot 10^{15}$	24216
530	2	$1.23 \cdot 10^{15}$	$5.20 \cdot 10^{15}$	20738
530	2	$1.04 \cdot 10^{15}$	$5.19 \cdot 10^{15}$	19422
530	2	$8.40 \cdot 10^{14}$	$5.19 \cdot 10^{15}$	15574
530	2	$6.50 \cdot 10^{14}$	$5.19 \cdot 10^{15}$	14994
530	2	$4.48 \cdot 10^{14}$	$5.20 \cdot 10^{15}$	11562
530	2	$2.24 \cdot 10^{14}$	$5.20 \cdot 10^{15}$	9379
529	2	0	$5.20 \cdot 10^{15}$	6906
529	5	0	$1.30 \cdot 10^{16}$	15285
529	5	$5.61 \cdot 10^{14}$	$1.30 \cdot 10^{16}$	24168
530	5	$1.04 \cdot 10^{15}$	$1.30 \cdot 10^{16}$	31365
530	5	$1.54 \cdot 10^{15}$	$1.30 \cdot 10^{16}$	35518
530	5	$2.02 \cdot 10^{15}$	$1.30 \cdot 10^{16}$	44140
529	5	$2.52 \cdot 10^{15}$	$1.30 \cdot 10^{16}$	47520
529	5	$3.00 \cdot 10^{15}$	$1.30 \cdot 10^{16}$	56046
529	5	$3.50 \cdot 10^{15}$	$1.30 \cdot 10^{16}$	64424
529	5	$3.98 \cdot 10^{15}$	$1.30 \cdot 10^{16}$	67818
529	5	$4.49 \cdot 10^{15}$	$1.30 \cdot 10^{16}$	77611
529	5	$5.05 \cdot 10^{15}$	$1.30 \cdot 10^{16}$	84487
569	5	$4.69 \cdot 10^{15}$	$1.21 \cdot 10^{16}$	85196
570	5	$4.25 \cdot 10^{15}$	$1.21 \cdot 10^{16}$	73282
570	5	$3.78 \cdot 10^{15}$	$1.21 \cdot 10^{16}$	61218
570	5	$3.33 \cdot 10^{15}$	$1.21 \cdot 10^{16}$	61411

T / K	P / bar	[DEE] / cm^{-3}	[HNO_3] / cm^{-3}	k'_3 / s^{-1}
570	5	$2.86 \cdot 10^{15}$	$1.21 \cdot 10^{16}$	54197
570	5	$2.42 \cdot 10^{15}$	$1.21 \cdot 10^{16}$	54934
570	5	$1.95 \cdot 10^{15}$	$1.21 \cdot 10^{16}$	43357
570	5	$1.51 \cdot 10^{15}$	$1.21 \cdot 10^{16}$	40216
570	5	$1.04 \cdot 10^{15}$	$1.21 \cdot 10^{16}$	31518
570	5	$5.21 \cdot 10^{14}$	$1.21 \cdot 10^{16}$	24653
570	5	0	$1.21 \cdot 10^{16}$	18019
570	2	0	$4.83 \cdot 10^{15}$	8582
569	2	$2.08 \cdot 10^{14}$	$4.83 \cdot 10^{15}$	9693
569	2	$3.86 \cdot 10^{14}$	$4.83 \cdot 10^{15}$	12521
569	2	$5.73 \cdot 10^{14}$	$4.83 \cdot 10^{15}$	16165
570	2	$7.50 \cdot 10^{14}$	$4.83 \cdot 10^{15}$	17735
570	2	$9.38 \cdot 10^{14}$	$4.83 \cdot 10^{15}$	19563
570	2	$1.11 \cdot 10^{15}$	$4.83 \cdot 10^{15}$	18372
570	2	$1.30 \cdot 10^{15}$	$4.83 \cdot 10^{15}$	20780
570	2	$1.48 \cdot 10^{15}$	$4.83 \cdot 10^{15}$	26221
570	2	$1.67 \cdot 10^{15}$	$4.83 \cdot 10^{15}$	25476
571	2	$1.87 \cdot 10^{15}$	$4.83 \cdot 10^{15}$	31809
489	10	$1.09 \cdot 10^{16}$	$2.94 \cdot 10^{16}$	201377
490	10	$9.88 \cdot 10^{15}$	$2.94 \cdot 10^{16}$	187788
490	10	$8.79 \cdot 10^{15}$	$2.94 \cdot 10^{16}$	165960
489	10	$7.76 \cdot 10^{15}$	$2.94 \cdot 10^{16}$	141364
488	10	$6.69 \cdot 10^{15}$	$2.95 \cdot 10^{16}$	139346
488	10	$5.65 \cdot 10^{15}$	$2.95 \cdot 10^{16}$	125971
488	10	$4.56 \cdot 10^{15}$	$2.95 \cdot 10^{16}$	104965
489	10	$3.52 \cdot 10^{15}$	$2.94 \cdot 10^{16}$	92036
489	10	$2.43 \cdot 10^{15}$	$2.94 \cdot 10^{16}$	69835
489	10	$1.21 \cdot 10^{15}$	$2.94 \cdot 10^{16}$	47901
489	10	0	$2.94 \cdot 10^{16}$	39979
530	10	0	$2.72 \cdot 10^{16}$	43802
529	10	$1.12 \cdot 10^{15}$	$2.72 \cdot 10^{16}$	60746
529	10	$2.08 \cdot 10^{15}$	$2.72 \cdot 10^{16}$	79214
529	10	$3.09 \cdot 10^{15}$	$2.72 \cdot 10^{16}$	96069
529	10	$4.04 \cdot 10^{15}$	$2.72 \cdot 10^{16}$	125987
529	10	$5.05 \cdot 10^{15}$	$2.72 \cdot 10^{16}$	145393

T / K	P / bar	[DEE] / cm^{-3}	[HNO_3] / cm^{-3}	k_3' / s^{-1}
529	10	$6.01 \cdot 10^{15}$	$2.72 \cdot 10^{16}$	145988
529	10	$7.02 \cdot 10^{15}$	$2.72 \cdot 10^{16}$	159489
529	10	$7.97 \cdot 10^{15}$	$2.72 \cdot 10^{16}$	169468
529	10	$8.98 \cdot 10^{15}$	$2.72 \cdot 10^{16}$	186158
529	10	$1.01 \cdot 10^{16}$	$2.72 \cdot 10^{16}$	192186
570	10	$9.37 \cdot 10^{15}$	$2.52 \cdot 10^{16}$	188696
570	10	$8.49 \cdot 10^{15}$	$2.52 \cdot 10^{16}$	171313
570	10	$7.55 \cdot 10^{15}$	$2.52 \cdot 10^{16}$	164348
570	10	$6.67 \cdot 10^{15}$	$2.53 \cdot 10^{16}$	160972
569	10	$5.73 \cdot 10^{15}$	$2.53 \cdot 10^{16}$	175631
569	10	$4.85 \cdot 10^{15}$	$2.53 \cdot 10^{16}$	140965
570	10	$3.91 \cdot 10^{15}$	$2.53 \cdot 10^{16}$	113512
570	10	$3.02 \cdot 10^{15}$	$2.53 \cdot 10^{16}$	116362
570	10	$2.08 \cdot 10^{15}$	$2.52 \cdot 10^{16}$	88642
570	10	$1.04 \cdot 10^{15}$	$2.52 \cdot 10^{16}$	73559
570	10	0	$2.52 \cdot 10^{16}$	68462

Table A.4: Overview of the experimental conditions in the measurements of the gas flow dependence of k'_3 in the system DEE + OH.

T / K	P / bar	f / slm	[DEE] / cm^{-3}	[HNO_3] / cm^{-3}	k'_3 / s^{-1}
294	10	0.7	$8.66 \cdot 10^{15}$	$4.82 \cdot 10^{16}$	248936
294	10	1.4	$8.66 \cdot 10^{15}$	$4.82 \cdot 10^{16}$	231004
294	10	2.1	$8.66 \cdot 10^{15}$	$4.82 \cdot 10^{16}$	236453
294	10	2.8	$8.66 \cdot 10^{15}$	$4.82 \cdot 10^{16}$	234301
294	10	4.9	$8.66 \cdot 10^{15}$	$4.82 \cdot 10^{16}$	238694
294	10	6.3	$8.66 \cdot 10^{15}$	$4.82 \cdot 10^{16}$	239734
294	10	7.7	$8.66 \cdot 10^{15}$	$4.82 \cdot 10^{16}$	223153
294	10	9.1	$8.66 \cdot 10^{15}$	$4.82 \cdot 10^{16}$	232721
570	2	0.7	$8.92 \cdot 10^{14}$	$5.19 \cdot 10^{15}$	12826
570	2	1.4	$8.92 \cdot 10^{14}$	$5.19 \cdot 10^{15}$	12195
571	2	2.1	$8.92 \cdot 10^{14}$	$5.19 \cdot 10^{15}$	16222
571	2	2.8	$8.92 \cdot 10^{14}$	$5.19 \cdot 10^{15}$	16587
570	2	3.5	$8.92 \cdot 10^{14}$	$5.19 \cdot 10^{15}$	21752
570	2	4.9	$8.92 \cdot 10^{14}$	$5.19 \cdot 10^{15}$	20235
570	2	6.3	$8.92 \cdot 10^{14}$	$5.19 \cdot 10^{15}$	20258
570	2	7.7	$8.92 \cdot 10^{14}$	$5.19 \cdot 10^{15}$	21246
570	2	9.1	$8.93 \cdot 10^{14}$	$5.20 \cdot 10^{15}$	19204

Table A.5: Overview of the experimental conditions in the measurements of the repetition rate dependence of k'_3 in the system DEE + OH.

T / K	P / bar	ν / Hz	[DEE] / cm^{-3}	[HNO_3] / cm^{-3}	k'_3 / s^{-1}
293	10	10	$8.68 \cdot 10^{15}$	$4.83 \cdot 10^{16}$	237965
293	10	5	$8.67 \cdot 10^{15}$	$4.83 \cdot 10^{16}$	216575
293	10	1	$8.67 \cdot 10^{15}$	$4.82 \cdot 10^{16}$	203962
570	2	10	$8.92 \cdot 10^{14}$	$5.19 \cdot 10^{15}$	21752
570	2	5	$8.93 \cdot 10^{14}$	$5.19 \cdot 10^{15}$	20165
571	2	1	$8.92 \cdot 10^{14}$	$5.19 \cdot 10^{15}$	20193

Table A.6: Overview of the experimental conditions in the measurements on the system DEE-d10 + OH and the obtained pseudo-first order rate coefficients k'_{3d}. The horizontal lines indicate the differentiation between the individual series of measurements.

T / K	P / bar	[DEE-d10] / cm^{-3}	[HNO_3] / cm^{-3}	k'_{3d} / s^{-1}
335	10	0	$1.51 \cdot 10^{16}$	41901
335	10	$5.49 \cdot 10^{15}$	$1.51 \cdot 10^{16}$	78118
335	10	$1.02 \cdot 10^{16}$	$1.51 \cdot 10^{16}$	106154
334	10	$1.51 \cdot 10^{16}$	$1.52 \cdot 10^{16}$	154585
334	10	$1.98 \cdot 10^{16}$	$1.52 \cdot 10^{16}$	162446
334	10	$2.47 \cdot 10^{16}$	$1.52 \cdot 10^{16}$	201548
334	10	$2.94 \cdot 10^{16}$	$1.52 \cdot 10^{16}$	227493
334	10	$3.44 \cdot 10^{16}$	$1.52 \cdot 10^{16}$	251390
334	10	$3.91 \cdot 10^{16}$	$1.52 \cdot 10^{16}$	278560
334	10	$4.40 \cdot 10^{16}$	$1.52 \cdot 10^{16}$	306958
334	10	$4.95 \cdot 10^{16}$	$1.52 \cdot 10^{16}$	308729
333	5	0	$7.60 \cdot 10^{15}$	15980
333	5	$2.75 \cdot 10^{15}$	$7.60 \cdot 10^{15}$	34494
333	5	$5.10 \cdot 10^{15}$	$7.60 \cdot 10^{15}$	53753
333	5	$7.58 \cdot 10^{15}$	$7.61 \cdot 10^{15}$	64604
333	5	$9.93 \cdot 10^{15}$	$7.61 \cdot 10^{15}$	76284
333	5	$1.24 \cdot 10^{16}$	$7.61 \cdot 10^{15}$	104010
333	5	$1.48 \cdot 10^{16}$	$7.61 \cdot 10^{15}$	120720
333	5	$1.72 \cdot 10^{16}$	$7.61 \cdot 10^{15}$	123127
333	5	$1.96 \cdot 10^{16}$	$7.61 \cdot 10^{15}$	134749
333	5	$2.21 \cdot 10^{16}$	$7.61 \cdot 10^{15}$	172767
333	5	$2.48 \cdot 10^{16}$	$7.61 \cdot 10^{15}$	135818
315	10	0	$1.61 \cdot 10^{16}$	58582
316	10	$5.81 \cdot 10^{15}$	$1.60 \cdot 10^{16}$	105797
316	10	$1.08 \cdot 10^{16}$	$1.60 \cdot 10^{16}$	129514
316	10	$1.60 \cdot 10^{16}$	$1.60 \cdot 10^{16}$	156102
316	10	$2.09 \cdot 10^{16}$	$1.60 \cdot 10^{16}$	189238
316	10	$2.61 \cdot 10^{16}$	$1.60 \cdot 10^{16}$	219449
316	10	$3.11 \cdot 10^{16}$	$1.60 \cdot 10^{16}$	238604
317	10	$3.63 \cdot 10^{16}$	$1.60 \cdot 10^{16}$	276099
317	10	$4.12 \cdot 10^{16}$	$1.60 \cdot 10^{16}$	267888

T / K	P / bar	[DEE-d10] / cm^{-3}	[HNO_3] / cm^{-3}	k'_{3d} / s^{-1}
317	10	$4.64 \cdot 10^{16}$	$1.60 \cdot 10^{16}$	374967
317	10	$5.22 \cdot 10^{16}$	$1.60 \cdot 10^{16}$	385154
317	5	$2.61 \cdot 10^{16}$	$8.00 \cdot 10^{15}$	180736
317	5	$2.32 \cdot 10^{16}$	$8.00 \cdot 10^{15}$	149157
317	5	$2.06 \cdot 10^{16}$	$7.99 \cdot 10^{15}$	146581
317	5	$1.81 \cdot 10^{16}$	$7.99 \cdot 10^{15}$	134134
317	5	$1.55 \cdot 10^{16}$	$7.99 \cdot 10^{15}$	101746
317	5	$1.30 \cdot 10^{16}$	$7.99 \cdot 10^{15}$	98922
317	5	$1.04 \cdot 10^{16}$	$7.99 \cdot 10^{15}$	93921
317	5	$7.96 \cdot 10^{15}$	$7.99 \cdot 10^{15}$	72340
317	5	$5.36 \cdot 10^{15}$	$7.99 \cdot 10^{15}$	59617
317	5	$2.89 \cdot 10^{15}$	$7.99 \cdot 10^{15}$	40363
317	5	0	$7.99 \cdot 10^{15}$	22651
394	5	0	$8.76 \cdot 10^{15}$	4857
394	5	$2.33 \cdot 10^{15}$	$8.75 \cdot 10^{15}$	18707
394	5	$4.31 \cdot 10^{15}$	$8.75 \cdot 10^{15}$	31555
394	5	$6.41 \cdot 10^{15}$	$8.75 \cdot 10^{15}$	45605
394	5	$8.39 \cdot 10^{15}$	$8.74 \cdot 10^{15}$	60529
394	5	$1.05 \cdot 10^{16}$	$8.74 \cdot 10^{15}$	69674
394	5	$1.25 \cdot 10^{16}$	$8.74 \cdot 10^{15}$	88146
394	5	$1.46 \cdot 10^{16}$	$8.74 \cdot 10^{15}$	105732
394	5	$1.65 \cdot 10^{16}$	$8.74 \cdot 10^{15}$	114227
394	5	$1.86 \cdot 10^{16}$	$8.74 \cdot 10^{15}$	127405
394	5	$2.10 \cdot 10^{16}$	$8.74 \cdot 10^{15}$	143648
395	10	$4.19 \cdot 10^{16}$	$1.75 \cdot 10^{16}$	345616
395	10	$3.72 \cdot 10^{16}$	$1.75 \cdot 10^{16}$	220370
395	10	$3.30 \cdot 10^{16}$	$1.75 \cdot 10^{16}$	215250
395	10	$2.91 \cdot 10^{16}$	$1.75 \cdot 10^{16}$	164478
395	10	$2.49 \cdot 10^{16}$	$1.74 \cdot 10^{16}$	161961
395	10	$2.09 \cdot 10^{16}$	$1.74 \cdot 10^{16}$	132166
395	10	$1.67 \cdot 10^{16}$	$1.74 \cdot 10^{16}$	98651
395	10	$1.28 \cdot 10^{16}$	$1.74 \cdot 10^{16}$	83151
395	10	$8.60 \cdot 10^{15}$	$1.74 \cdot 10^{16}$	56010
395	10	$4.65 \cdot 10^{15}$	$1.74 \cdot 10^{16}$	32490
395	10	0	$1.74 \cdot 10^{16}$	9512

T / K	P / bar	[DEE-d10] / cm^{-3}	[HNO_3] / cm^{-3}	k'_{3d} / s^{-1}
395	2	$8.36 \cdot 10^{15}$	$3.49 \cdot 10^{15}$	57078
395	2	$7.43 \cdot 10^{15}$	$3.49 \cdot 10^{15}$	49796
395	2	$6.59 \cdot 10^{15}$	$3.49 \cdot 10^{15}$	51540
396	2	$5.80 \cdot 10^{15}$	$3.49 \cdot 10^{15}$	40472
396	2	$4.97 \cdot 10^{15}$	$3.49 \cdot 10^{15}$	34070
396	2	$4.18 \cdot 10^{15}$	$3.48 \cdot 10^{15}$	30045
396	2	$3.34 \cdot 10^{15}$	$3.48 \cdot 10^{15}$	24782
396	2	$2.55 \cdot 10^{15}$	$3.48 \cdot 10^{15}$	19625
396	2	$1.72 \cdot 10^{15}$	$3.48 \cdot 10^{15}$	13216
396	2	$9.28 \cdot 10^{14}$	$3.48 \cdot 10^{15}$	7366
396	2	0	$3.48 \cdot 10^{15}$	2137
424	10	0	$1.63 \cdot 10^{16}$	12651
424	10	$4.33 \cdot 10^{15}$	$1.63 \cdot 10^{16}$	40338
424	10	$8.01 \cdot 10^{15}$	$1.63 \cdot 10^{16}$	70399
424	10	$1.19 \cdot 10^{16}$	$1.63 \cdot 10^{16}$	93221
424	10	$1.56 \cdot 10^{16}$	$1.62 \cdot 10^{16}$	117380
424	10	$1.95 \cdot 10^{16}$	$1.62 \cdot 10^{16}$	144666
424	10	$2.32 \cdot 10^{16}$	$1.62 \cdot 10^{16}$	171626
424	10	$2.71 \cdot 10^{16}$	$1.62 \cdot 10^{16}$	215512
424	10	$3.07 \cdot 10^{16}$	$1.62 \cdot 10^{16}$	235517
424	10	$3.46 \cdot 10^{16}$	$1.62 \cdot 10^{16}$	268337
424	10	$3.90 \cdot 10^{16}$	$1.62 \cdot 10^{16}$	313232
424	5	$1.95 \cdot 10^{16}$	$8.13 \cdot 10^{15}$	143759
424	5	$1.73 \cdot 10^{16}$	$8.12 \cdot 10^{15}$	122398
424	5	$1.54 \cdot 10^{16}$	$8.12 \cdot 10^{15}$	114474
424	5	$1.35 \cdot 10^{16}$	$8.12 \cdot 10^{15}$	111319
424	5	$1.16 \cdot 10^{16}$	$8.12 \cdot 10^{15}$	88946
424	5	$9.74 \cdot 10^{15}$	$8.12 \cdot 10^{15}$	74047
424	5	$7.79 \cdot 10^{15}$	$8.12 \cdot 10^{15}$	62276
424	5	$5.95 \cdot 10^{15}$	$8.12 \cdot 10^{15}$	44359
424	5	$4.00 \cdot 10^{15}$	$8.12 \cdot 10^{15}$	29504
424	5	$2.16 \cdot 10^{15}$	$8.12 \cdot 10^{15}$	17941
424	5	0	$8.12 \cdot 10^{15}$	4662
449	5	0	$7.67 \cdot 10^{15}$	7967
450	5	$2.04 \cdot 10^{15}$	$7.66 \cdot 10^{15}$	21869

T / K	P / bar	[DEE-d10] / cm^{-3}	[HNO_3] / cm^{-3}	k'_{3d} / s^{-1}
450	5	$3.78 \cdot 10^{15}$	$7.66 \cdot 10^{15}$	32175
450	5	$5.61 \cdot 10^{15}$	$7.66 \cdot 10^{15}$	48752
450	5	$7.35 \cdot 10^{15}$	$7.66 \cdot 10^{15}$	57277
450	5	$9.18 \cdot 10^{15}$	$7.66 \cdot 10^{15}$	73248
450	5	$1.09 \cdot 10^{16}$	$7.66 \cdot 10^{15}$	92721
450	5	$1.28 \cdot 10^{16}$	$7.66 \cdot 10^{15}$	101735
450	5	$1.45 \cdot 10^{16}$	$7.66 \cdot 10^{15}$	111576
450	5	$1.63 \cdot 10^{16}$	$7.65 \cdot 10^{15}$	120764
450	5	$1.84 \cdot 10^{16}$	$7.65 \cdot 10^{15}$	134610
485	5	$1.70 \cdot 10^{16}$	$7.11 \cdot 10^{15}$	134907
485	5	$1.51 \cdot 10^{16}$	$7.10 \cdot 10^{15}$	145046
485	5	$1.34 \cdot 10^{16}$	$7.10 \cdot 10^{15}$	127524
485	5	$1.18 \cdot 10^{16}$	$7.10 \cdot 10^{15}$	112342
485	5	$1.01 \cdot 10^{16}$	$7.10 \cdot 10^{15}$	90765
485	5	$8.51 \cdot 10^{15}$	$7.10 \cdot 10^{15}$	81817
485	5	$6.81 \cdot 10^{15}$	$7.10 \cdot 10^{15}$	64889
485	5	$5.20 \cdot 10^{15}$	$7.10 \cdot 10^{15}$	56005
485	5	$3.50 \cdot 10^{15}$	$7.10 \cdot 10^{15}$	39329
485	5	$1.89 \cdot 10^{15}$	$7.10 \cdot 10^{15}$	27003
485	5	0	$7.10 \cdot 10^{15}$	12017
372	5	0	$1.94 \cdot 10^{16}$	8761
373	5	$2.46 \cdot 10^{15}$	$1.94 \cdot 10^{16}$	25186
373	5	$4.56 \cdot 10^{15}$	$1.94 \cdot 10^{16}$	38429
373	5	$6.78 \cdot 10^{15}$	$1.94 \cdot 10^{16}$	54912
373	5	$8.87 \cdot 10^{15}$	$1.94 \cdot 10^{16}$	65718
373	5	$1.11 \cdot 10^{16}$	$1.94 \cdot 10^{16}$	80286
373	5	$1.32 \cdot 10^{16}$	$1.94 \cdot 10^{16}$	91984
373	5	$1.54 \cdot 10^{16}$	$1.94 \cdot 10^{16}$	110659
373	5	$1.75 \cdot 10^{16}$	$1.94 \cdot 10^{16}$	119746
373	5	$1.97 \cdot 10^{16}$	$1.94 \cdot 10^{16}$	139883
373	5	$2.22 \cdot 10^{16}$	$1.94 \cdot 10^{16}$	150966
373	10	$4.44 \cdot 10^{16}$	$3.87 \cdot 10^{16}$	294807
373	10	$3.94 \cdot 10^{16}$	$3.87 \cdot 10^{16}$	262281
373	10	$3.50 \cdot 10^{16}$	$3.87 \cdot 10^{16}$	243855
373	10	$3.08 \cdot 10^{16}$	$3.87 \cdot 10^{16}$	206913

T / K	P / bar	[DEE-d10] / cm^{-3}	[HNO_3] / cm^{-3}	k'_{3d} / s^{-1}
373	10	$2.64 \cdot 10^{16}$	$3.87 \cdot 10^{16}$	178355
373	10	$2.22 \cdot 10^{16}$	$3.87 \cdot 10^{16}$	147453
373	10	$1.77 \cdot 10^{16}$	$3.87 \cdot 10^{16}$	112354
373	10	$1.36 \cdot 10^{16}$	$3.87 \cdot 10^{16}$	97588
373	10	$9.12 \cdot 10^{15}$	$3.87 \cdot 10^{16}$	65421
373	10	$4.93 \cdot 10^{15}$	$3.87 \cdot 10^{16}$	41006
373	10	0	$3.87 \cdot 10^{16}$	17171
373	2	0	$7.75 \cdot 10^{15}$	3574
373	2	$9.84 \cdot 10^{14}$	$7.74 \cdot 10^{15}$	9128
373	2	$1.82 \cdot 10^{15}$	$7.74 \cdot 10^{15}$	14428
373	2	$2.71 \cdot 10^{15}$	$7.74 \cdot 10^{15}$	21074
373	2	$3.55 \cdot 10^{15}$	$7.74 \cdot 10^{15}$	25745
373	2	$4.43 \cdot 10^{15}$	$7.74 \cdot 10^{15}$	33266
373	2	$5.27 \cdot 10^{15}$	$7.74 \cdot 10^{15}$	37860
373	2	$6.16 \cdot 10^{15}$	$7.74 \cdot 10^{15}$	43502
373	2	$6.99 \cdot 10^{15}$	$7.74 \cdot 10^{15}$	47821
373	2	$7.88 \cdot 10^{15}$	$7.74 \cdot 10^{15}$	57594
373	2	$8.86 \cdot 10^{15}$	$7.74 \cdot 10^{15}$	61550
355	2	$9.31 \cdot 10^{15}$	$8.13 \cdot 10^{15}$	68002
355	2	$8.28 \cdot 10^{15}$	$8.13 \cdot 10^{15}$	60782
355	2	$7.34 \cdot 10^{15}$	$8.13 \cdot 10^{15}$	48606
355	2	$6.46 \cdot 10^{15}$	$8.13 \cdot 10^{15}$	46191
355	2	$5.53 \cdot 10^{15}$	$8.12 \cdot 10^{15}$	41638
355	2	$4.65 \cdot 10^{15}$	$8.12 \cdot 10^{15}$	33076
355	2	$3.72 \cdot 10^{15}$	$8.12 \cdot 10^{15}$	27317
356	2	$2.84 \cdot 10^{15}$	$8.12 \cdot 10^{15}$	21515
356	2	$1.91 \cdot 10^{15}$	$8.12 \cdot 10^{15}$	15330
356	2	$1.03 \cdot 10^{15}$	$8.12 \cdot 10^{15}$	9703
356	2	0	$8.12 \cdot 10^{15}$	3776
356	5	0	$2.03 \cdot 10^{16}$	8761
356	5	$2.58 \cdot 10^{15}$	$2.03 \cdot 10^{16}$	23591
356	5	$4.77 \cdot 10^{15}$	$2.03 \cdot 10^{16}$	37648
356	5	$7.10 \cdot 10^{15}$	$2.03 \cdot 10^{16}$	49791
356	5	$9.29 \cdot 10^{15}$	$2.03 \cdot 10^{16}$	62871
356	5	$1.16 \cdot 10^{16}$	$2.03 \cdot 10^{16}$	83959

T / K	P / bar	[DEE-d10] / cm^{-3}	[HNO_3] / cm^{-3}	k'_{3d} / s^{-1}
356	5	$1.38 \cdot 10^{16}$	$2.03 \cdot 10^{16}$	100962
356	5	$1.61 \cdot 10^{16}$	$2.03 \cdot 10^{16}$	113293
356	5	$1.83 \cdot 10^{16}$	$2.03 \cdot 10^{16}$	136938
356	5	$2.06 \cdot 10^{16}$	$2.03 \cdot 10^{16}$	146359
356	5	$2.32 \cdot 10^{16}$	$2.03 \cdot 10^{16}$	170870
356	10	$4.65 \cdot 10^{16}$	$4.06 \cdot 10^{16}$	341366
356	10	$4.13 \cdot 10^{16}$	$4.06 \cdot 10^{16}$	291201
356	10	$3.66 \cdot 10^{16}$	$4.06 \cdot 10^{16}$	269320
356	10	$3.23 \cdot 10^{16}$	$4.06 \cdot 10^{16}$	228476
356	10	$2.76 \cdot 10^{16}$	$4.06 \cdot 10^{16}$	197438
356	10	$2.32 \cdot 10^{16}$	$4.06 \cdot 10^{16}$	168609
356	10	$1.86 \cdot 10^{16}$	$4.06 \cdot 10^{16}$	136826
356	10	$1.42 \cdot 10^{16}$	$4.06 \cdot 10^{16}$	106751
356	10	$9.55 \cdot 10^{15}$	$4.06 \cdot 10^{16}$	75047
356	10	$5.16 \cdot 10^{15}$	$4.06 \cdot 10^{16}$	53387
356	10	0	$4.06 \cdot 10^{16}$	22697
294	10	0	$4.90 \cdot 10^{16}$	65328
295	10	$6.24 \cdot 10^{15}$	$4.90 \cdot 10^{16}$	94587
295	10	$1.25 \cdot 10^{16}$	$4.90 \cdot 10^{16}$	135955
295	10	$1.71 \cdot 10^{16}$	$4.90 \cdot 10^{16}$	164023
295	10	$2.24 \cdot 10^{16}$	$4.90 \cdot 10^{16}$	198528
295	10	$2.80 \cdot 10^{16}$	$4.90 \cdot 10^{16}$	235710
295	10	$3.33 \cdot 10^{16}$	$4.90 \cdot 10^{16}$	276080
295	10	$3.89 \cdot 10^{16}$	$4.90 \cdot 10^{16}$	301775
295	10	$4.42 \cdot 10^{16}$	$4.90 \cdot 10^{16}$	323803
295	10	$4.98 \cdot 10^{16}$	$4.89 \cdot 10^{16}$	342863
295	10	$5.60 \cdot 10^{16}$	$4.89 \cdot 10^{16}$	400454
295	5	$2.80 \cdot 10^{16}$	$2.45 \cdot 10^{16}$	196163
295	5	$2.49 \cdot 10^{16}$	$2.45 \cdot 10^{16}$	179976
295	5	$2.21 \cdot 10^{16}$	$2.45 \cdot 10^{16}$	175345
295	5	$1.94 \cdot 10^{16}$	$2.44 \cdot 10^{16}$	134454
295	5	$1.66 \cdot 10^{16}$	$2.44 \cdot 10^{16}$	126671
295	5	$1.40 \cdot 10^{16}$	$2.44 \cdot 10^{16}$	109736
295	5	$1.12 \cdot 10^{16}$	$2.44 \cdot 10^{16}$	93573
295	5	$8.55 \cdot 10^{15}$	$2.44 \cdot 10^{16}$	75842

T / K	P / bar	[DEE-d10] / cm^{-3}	[HNO_3] / cm^{-3}	k'_{3d} / s^{-1}
295	5	$5.75 \cdot 10^{15}$	$2.44 \cdot 10^{16}$	62126
295	5	$3.11 \cdot 10^{15}$	$2.44 \cdot 10^{16}$	43188
296	5	0	$2.44 \cdot 10^{16}$	26886
296	2	0	$9.77 \cdot 10^{15}$	8032
296	2	$1.24 \cdot 10^{15}$	$9.77 \cdot 10^{15}$	14362
296	2	$2.30 \cdot 10^{15}$	$9.76 \cdot 10^{15}$	21415
296	2	$3.42 \cdot 10^{15}$	$9.76 \cdot 10^{15}$	29252
296	2	$4.47 \cdot 10^{15}$	$9.76 \cdot 10^{15}$	36562
296	2	$5.59 \cdot 10^{15}$	$9.76 \cdot 10^{15}$	45070
296	2	$6.64 \cdot 10^{15}$	$9.76 \cdot 10^{15}$	48699
296	2	$7.76 \cdot 10^{15}$	$9.76 \cdot 10^{15}$	59281
296	2	$8.81 \cdot 10^{15}$	$9.76 \cdot 10^{15}$	64091
296	2	$9.93 \cdot 10^{15}$	$9.76 \cdot 10^{15}$	74987
296	2	$1.12 \cdot 10^{16}$	$9.76 \cdot 10^{15}$	77178
300	10	$5.50 \cdot 10^{16}$	$4.81 \cdot 10^{16}$	427578
300	10	$4.89 \cdot 10^{16}$	$4.81 \cdot 10^{16}$	379585
300	10	$4.34 \cdot 10^{16}$	$4.80 \cdot 10^{16}$	350045
301	10	$3.82 \cdot 10^{16}$	$4.80 \cdot 10^{16}$	309047
301	10	$3.27 \cdot 10^{16}$	$4.80 \cdot 10^{16}$	273854
301	10	$2.75 \cdot 10^{16}$	$4.80 \cdot 10^{16}$	262078
301	10	$2.20 \cdot 10^{16}$	$4.80 \cdot 10^{16}$	193410
301	10	$1.68 \cdot 10^{16}$	$4.80 \cdot 10^{16}$	177614
301	10	$1.13 \cdot 10^{16}$	$4.80 \cdot 10^{16}$	140857
301	10	$6.11 \cdot 10^{15}$	$4.80 \cdot 10^{16}$	106248
301	10	0	$4.80 \cdot 10^{16}$	69306
301	5	0	$2.40 \cdot 10^{16}$	26350
301	5	$3.05 \cdot 10^{15}$	$2.40 \cdot 10^{16}$	44535
301	5	$5.64 \cdot 10^{15}$	$2.40 \cdot 10^{16}$	64082
301	5	$8.39 \cdot 10^{15}$	$2.40 \cdot 10^{16}$	85432
301	5	$1.10 \cdot 10^{16}$	$2.40 \cdot 10^{16}$	97936
301	5	$1.37 \cdot 10^{16}$	$2.40 \cdot 10^{16}$	108805
301	5	$1.63 \cdot 10^{16}$	$2.40 \cdot 10^{16}$	135781
301	5	$1.91 \cdot 10^{16}$	$2.40 \cdot 10^{16}$	142394
301	5	$2.16 \cdot 10^{16}$	$2.40 \cdot 10^{16}$	164884
301	5	$2.44 \cdot 10^{16}$	$2.40 \cdot 10^{16}$	167711

T / K	P / bar	[DEE-d10] / cm^{-3}	[HNO_3] / cm^{-3}	k'_{3d} / s^{-1}
301	5	$2.74 \cdot 10^{16}$	$2.40 \cdot 10^{16}$	213369
301	2	$1.10 \cdot 10^{16}$	$9.58 \cdot 10^{15}$	73150
301	2	$9.75 \cdot 10^{15}$	$9.58 \cdot 10^{15}$	74866
301	2	$8.65 \cdot 10^{15}$	$9.58 \cdot 10^{15}$	60146
301	2	$7.62 \cdot 10^{15}$	$9.58 \cdot 10^{15}$	63628
301	2	$6.52 \cdot 10^{15}$	$9.58 \cdot 10^{15}$	49056
302	2	$5.48 \cdot 10^{15}$	$9.58 \cdot 10^{15}$	40040
302	2	$4.39 \cdot 10^{15}$	$9.57 \cdot 10^{15}$	34855
302	2	$3.35 \cdot 10^{15}$	$9.57 \cdot 10^{15}$	29444
302	2	$2.25 \cdot 10^{15}$	$9.57 \cdot 10^{15}$	20963
302	2	$1.22 \cdot 10^{15}$	$9.57 \cdot 10^{15}$	15101
302	2	0	$9.57 \cdot 10^{15}$	7718
317	2	0	$9.09 \cdot 10^{15}$	6733
318	2	$1.16 \cdot 10^{15}$	$9.09 \cdot 10^{15}$	13737
318	2	$2.14 \cdot 10^{15}$	$9.08 \cdot 10^{15}$	19603
318	2	$3.18 \cdot 10^{15}$	$9.08 \cdot 10^{15}$	27389
318	2	$4.16 \cdot 10^{15}$	$9.08 \cdot 10^{15}$	34393
318	2	$5.20 \cdot 10^{15}$	$9.07 \cdot 10^{15}$	41541
318	2	$6.17 \cdot 10^{15}$	$9.07 \cdot 10^{15}$	44826
318	2	$7.21 \cdot 10^{15}$	$9.07 \cdot 10^{15}$	52750
318	2	$8.19 \cdot 10^{15}$	$9.07 \cdot 10^{15}$	61789
319	2	$9.23 \cdot 10^{15}$	$9.06 \cdot 10^{15}$	68114
319	2	$1.04 \cdot 10^{16}$	$9.06 \cdot 10^{15}$	73875
334	2	$9.89 \cdot 10^{15}$	$8.63 \cdot 10^{15}$	70811
334	2	$8.79 \cdot 10^{15}$	$8.63 \cdot 10^{15}$	63419
334	2	$7.80 \cdot 10^{15}$	$8.63 \cdot 10^{15}$	56614
335	2	$6.86 \cdot 10^{15}$	$8.63 \cdot 10^{15}$	50821
335	2	$5.87 \cdot 10^{15}$	$8.63 \cdot 10^{15}$	45270
335	2	$4.94 \cdot 10^{15}$	$8.63 \cdot 10^{15}$	38943
335	2	$3.95 \cdot 10^{15}$	$8.63 \cdot 10^{15}$	31419
335	2	$3.02 \cdot 10^{15}$	$8.63 \cdot 10^{15}$	26705
335	2	$2.03 \cdot 10^{15}$	$8.63 \cdot 10^{15}$	17949
335	2	$1.10 \cdot 10^{15}$	$8.63 \cdot 10^{15}$	13028
335	2	0	$8.63 \cdot 10^{15}$	6037
452	10	0	$3.27 \cdot 10^{16}$	28186

T / K	P / bar	[DEE-d10] / cm^{-3}	[HNO_3] / cm^{-3}	k'_{3d} / s^{-1}
453	10	$4.06 \cdot 10^{15}$	$3.27 \cdot 10^{16}$	64871
453	10	$7.51 \cdot 10^{15}$	$3.27 \cdot 10^{16}$	93761
453	10	$1.12 \cdot 10^{16}$	$3.27 \cdot 10^{16}$	120963
453	10	$1.46 \cdot 10^{16}$	$3.27 \cdot 10^{16}$	160133
453	10	$1.83 \cdot 10^{16}$	$3.27 \cdot 10^{16}$	179199
453	10	$2.17 \cdot 10^{16}$	$3.27 \cdot 10^{16}$	196493
453	10	$2.53 \cdot 10^{16}$	$3.27 \cdot 10^{16}$	233929
453	10	$2.88 \cdot 10^{16}$	$3.27 \cdot 10^{16}$	228936
453	10	$3.24 \cdot 10^{16}$	$3.27 \cdot 10^{16}$	303051
453	10	$3.65 \cdot 10^{16}$	$3.27 \cdot 10^{16}$	306294
428	2	0	$6.90 \cdot 10^{15}$	3679
429	2	$8.57 \cdot 10^{14}$	$6.90 \cdot 10^{15}$	11084
429	2	$1.59 \cdot 10^{15}$	$6.90 \cdot 10^{15}$	14181
429	2	$2.36 \cdot 10^{15}$	$6.90 \cdot 10^{15}$	18008
429	2	$3.08 \cdot 10^{15}$	$6.90 \cdot 10^{15}$	26412
429	2	$3.86 \cdot 10^{15}$	$6.90 \cdot 10^{15}$	30890
429	2	$4.58 \cdot 10^{15}$	$6.90 \cdot 10^{15}$	34208
429	2	$5.35 \cdot 10^{15}$	$6.90 \cdot 10^{15}$	36567
429	2	$6.08 \cdot 10^{15}$	$6.90 \cdot 10^{15}$	40586
429	2	$6.85 \cdot 10^{15}$	$6.89 \cdot 10^{15}$	46710
429	2	$7.70 \cdot 10^{15}$	$6.89 \cdot 10^{15}$	55054
488	2	$4.04 \cdot 10^{15}$	$5.81 \cdot 10^{15}$	40398
488	2	$3.59 \cdot 10^{15}$	$5.81 \cdot 10^{15}$	38859
488	2	$3.19 \cdot 10^{15}$	$5.81 \cdot 10^{15}$	37220
488	2	$2.81 \cdot 10^{15}$	$5.81 \cdot 10^{15}$	31307
488	2	$2.40 \cdot 10^{15}$	$5.81 \cdot 10^{15}$	30678
488	2	$2.02 \cdot 10^{15}$	$5.81 \cdot 10^{15}$	27852
488	2	$1.62 \cdot 10^{15}$	$5.81 \cdot 10^{15}$	21378
488	2	$1.23 \cdot 10^{15}$	$5.81 \cdot 10^{15}$	18678
488	2	$8.30 \cdot 10^{14}$	$5.81 \cdot 10^{15}$	17584
488	2	$4.49 \cdot 10^{14}$	$5.81 \cdot 10^{15}$	13343
488	2	0	$5.81 \cdot 10^{15}$	8566
450	2	0	$6.39 \cdot 10^{15}$	5614
451	2	$4.87 \cdot 10^{14}$	$6.38 \cdot 10^{15}$	9673
451	2	$9.00 \cdot 10^{14}$	$6.38 \cdot 10^{15}$	12234

T / K	P / bar	[DEE-d10] / cm^{-3}	[HNO_3] / cm^{-3}	k'_{3d} / s^{-1}
450	2	$1.10 \cdot 10^{15}$	$6.39 \cdot 10^{15}$	15480
450	2	$1.75 \cdot 10^{15}$	$6.39 \cdot 10^{15}$	17707
450	2	$2.19 \cdot 10^{15}$	$6.39 \cdot 10^{15}$	22315
451	2	$2.60 \cdot 10^{15}$	$6.39 \cdot 10^{15}$	24568
451	2	$3.04 \cdot 10^{15}$	$6.38 \cdot 10^{15}$	26574
451	2	$3.46 \cdot 10^{15}$	$6.39 \cdot 10^{15}$	28322
451	2	$3.89 \cdot 10^{15}$	$6.38 \cdot 10^{15}$	31287
451	2	$4.38 \cdot 10^{15}$	$6.38 \cdot 10^{15}$	34260
486	10	$2.03 \cdot 10^{16}$	$2.96 \cdot 10^{16}$	200076
486	10	$1.80 \cdot 10^{16}$	$2.96 \cdot 10^{16}$	201797
486	10	$1.60 \cdot 10^{16}$	$2.96 \cdot 10^{16}$	174654
487	10	$1.41 \cdot 10^{16}$	$2.96 \cdot 10^{16}$	150914
487	10	$1.21 \cdot 10^{16}$	$2.96 \cdot 10^{16}$	136280
487	10	$1.01 \cdot 10^{16}$	$2.96 \cdot 10^{16}$	122833
487	10	$8.11 \cdot 10^{15}$	$2.96 \cdot 10^{16}$	102129
487	10	$6.20 \cdot 10^{15}$	$2.96 \cdot 10^{16}$	90841
486	10	$4.17 \cdot 10^{15}$	$2.96 \cdot 10^{16}$	71756
486	10	$2.25 \cdot 10^{15}$	$2.96 \cdot 10^{16}$	57697
487	10	0	$2.96 \cdot 10^{16}$	39291
529	10	0	$2.72 \cdot 10^{16}$	52434
529	10	$2.07 \cdot 10^{15}$	$2.72 \cdot 10^{16}$	72417
529	10	$3.83 \cdot 10^{15}$	$2.72 \cdot 10^{16}$	95485
529	10	$4.66 \cdot 10^{15}$	$2.72 \cdot 10^{16}$	110125
529	10	$7.46 \cdot 10^{15}$	$2.72 \cdot 10^{16}$	116275
529	10	$9.32 \cdot 10^{15}$	$2.72 \cdot 10^{16}$	151540
529	10	$1.11 \cdot 10^{16}$	$2.72 \cdot 10^{16}$	141470
529	10	$1.30 \cdot 10^{16}$	$2.72 \cdot 10^{16}$	165560
529	10	$1.47 \cdot 10^{16}$	$2.72 \cdot 10^{16}$	205634
530	10	$1.66 \cdot 10^{16}$	$2.72 \cdot 10^{16}$	194629
530	10	$1.86 \cdot 10^{16}$	$2.72 \cdot 10^{16}$	195189
530	5	$9.31 \cdot 10^{15}$	$1.36 \cdot 10^{16}$	90340
530	5	$8.28 \cdot 10^{15}$	$1.36 \cdot 10^{16}$	82668
530	5	$7.35 \cdot 10^{15}$	$1.36 \cdot 10^{16}$	76422
529	5	$6.47 \cdot 10^{15}$	$1.36 \cdot 10^{16}$	66831
529	5	$5.54 \cdot 10^{15}$	$1.36 \cdot 10^{16}$	59535

T / K	P / bar	[DEE-d10] / cm^{-3}	[HNO_3] / cm^{-3}	k'_{3d} / s^{-1}
530	5	$4.66 \cdot 10^{15}$	$1.36 \cdot 10^{16}$	55878
529	5	$3.73 \cdot 10^{15}$	$1.36 \cdot 10^{16}$	47350
529	5	$2.85 \cdot 10^{15}$	$1.36 \cdot 10^{16}$	41995
529	5	$1.92 \cdot 10^{15}$	$1.36 \cdot 10^{16}$	33600
530	5	$1.04 \cdot 10^{15}$	$1.36 \cdot 10^{16}$	24029
530	5	0	$1.36 \cdot 10^{16}$	18793
529	2	0	$5.44 \cdot 10^{15}$	7078
529	2	$4.14 \cdot 10^{14}$	$5.44 \cdot 10^{15}$	10844
529	2	$7.66 \cdot 10^{14}$	$5.44 \cdot 10^{15}$	13885
530	2	$9.32 \cdot 10^{14}$	$5.43 \cdot 10^{15}$	15936
530	2	$1.49 \cdot 10^{15}$	$5.43 \cdot 10^{15}$	18240
529	2	$1.86 \cdot 10^{15}$	$5.44 \cdot 10^{15}$	19686
530	2	$2.21 \cdot 10^{15}$	$5.43 \cdot 10^{15}$	25869
530	2	$2.59 \cdot 10^{15}$	$5.43 \cdot 10^{15}$	29157
530	2	$2.94 \cdot 10^{15}$	$5.43 \cdot 10^{15}$	26663
530	2	$3.31 \cdot 10^{15}$	$5.43 \cdot 10^{15}$	34320
529	2	$3.73 \cdot 10^{15}$	$5.44 \cdot 10^{15}$	35487
567	2	$3.48 \cdot 10^{15}$	$5.08 \cdot 10^{15}$	36168
567	2	$3.09 \cdot 10^{15}$	$5.07 \cdot 10^{15}$	35630
568	2	$2.74 \cdot 10^{15}$	$5.07 \cdot 10^{15}$	31713
568	2	$2.41 \cdot 10^{15}$	$5.07 \cdot 10^{15}$	25643
568	2	$2.07 \cdot 10^{15}$	$5.07 \cdot 10^{15}$	23119
567	2	$1.74 \cdot 10^{15}$	$5.07 \cdot 10^{15}$	15606
567	2	$1.39 \cdot 10^{15}$	$5.07 \cdot 10^{15}$	22514
567	2	$1.06 \cdot 10^{15}$	$5.07 \cdot 10^{15}$	18163
567	2	$7.15 \cdot 10^{14}$	$5.07 \cdot 10^{15}$	14675
567	2	$3.86 \cdot 10^{14}$	$5.07 \cdot 10^{15}$	11247
568	2	0	$5.07 \cdot 10^{15}$	8852
568	5	0	$1.27 \cdot 10^{16}$	21725
568	5	$9.65 \cdot 10^{14}$	$1.27 \cdot 10^{16}$	29180
568	5	$1.79 \cdot 10^{15}$	$1.27 \cdot 10^{16}$	30296
567	5	$2.18 \cdot 10^{15}$	$1.27 \cdot 10^{16}$	45265
567	5	$3.48 \cdot 10^{15}$	$1.27 \cdot 10^{16}$	49706
567	5	$4.35 \cdot 10^{15}$	$1.27 \cdot 10^{16}$	62439
567	5	$5.17 \cdot 10^{15}$	$1.27 \cdot 10^{16}$	65014

T / K	P / bar	[DEE-d10] / cm^{-3}	[HNO_3] / cm^{-3}	k'_{3d} / s^{-1}
568	5	$6.04 \cdot 10^{15}$	$1.27 \cdot 10^{16}$	72878
568	5	$6.85 \cdot 10^{15}$	$1.27 \cdot 10^{16}$	87212
568	5	$7.72 \cdot 10^{15}$	$1.27 \cdot 10^{16}$	70501
568	5	$8.69 \cdot 10^{15}$	$1.27 \cdot 10^{16}$	87999
567	10	$1.74 \cdot 10^{16}$	$2.54 \cdot 10^{16}$	212986
568	10	$1.55 \cdot 10^{16}$	$2.53 \cdot 10^{16}$	209313
568	10	$1.37 \cdot 10^{16}$	$2.53 \cdot 10^{16}$	195435
568	10	$1.21 \cdot 10^{16}$	$2.53 \cdot 10^{16}$	130939
567	10	$1.03 \cdot 10^{16}$	$2.54 \cdot 10^{16}$	167968
567	10	$8.70 \cdot 10^{15}$	$2.54 \cdot 10^{16}$	130723
567	10	$6.96 \cdot 10^{15}$	$2.54 \cdot 10^{16}$	124268
567	10	$5.32 \cdot 10^{15}$	$2.54 \cdot 10^{16}$	103096
567	10	$3.58 \cdot 10^{15}$	$2.54 \cdot 10^{16}$	87416
568	10	$1.93 \cdot 10^{15}$	$2.53 \cdot 10^{16}$	68259
568	10	0	$2.53 \cdot 10^{16}$	46235

A.4 Appendix to Chapter 7

Table A.7: Overview of the experimental conditions in the measurements on the system DMM + OH and the obtained pseudo-first order rate coefficients k'_4. The horizontal lines indicate the differentiation between the individual series of measurements.

T / K	P / bar	[DMM] / cm^{-3}	[HNO_3] / cm^{-3}	k'_4 / s^{-1}
297	10	0	$4.95{\cdot}10^{16}$	29813
297	10	$5.28{\cdot}10^{15}$	$4.95{\cdot}10^{16}$	55243
297	10	$1.06{\cdot}10^{16}$	$4.95{\cdot}10^{16}$	80682
297	10	$1.58{\cdot}10^{16}$	$4.94{\cdot}10^{16}$	110953
297	10	$2.11{\cdot}10^{16}$	$4.94{\cdot}10^{16}$	138571
297	10	$2.64{\cdot}10^{16}$	$4.94{\cdot}10^{16}$	163288
297	10	$3.17{\cdot}10^{16}$	$4.94{\cdot}10^{16}$	199079
297	10	$3.70{\cdot}10^{16}$	$4.94{\cdot}10^{16}$	207464
297	10	$4.22{\cdot}10^{16}$	$4.94{\cdot}10^{16}$	241027
297	10	$4.75{\cdot}10^{16}$	$4.94{\cdot}10^{16}$	268574
297	10	$5.28{\cdot}10^{16}$	$4.94{\cdot}10^{16}$	290207
297	5	0	$2.47{\cdot}10^{16}$	12369
297	5	$2.64{\cdot}10^{15}$	$2.47{\cdot}10^{16}$	23262
297	5	$5.27{\cdot}10^{15}$	$2.47{\cdot}10^{16}$	36595
297	5	$7.91{\cdot}10^{15}$	$2.47{\cdot}10^{16}$	48412
297	5	$1.05{\cdot}10^{16}$	$2.47{\cdot}10^{16}$	59889
297	5	$1.32{\cdot}10^{16}$	$2.47{\cdot}10^{16}$	70395
297	5	$1.58{\cdot}10^{16}$	$2.47{\cdot}10^{16}$	84505
297	5	$1.84{\cdot}10^{16}$	$2.47{\cdot}10^{16}$	95309
297	5	$2.11{\cdot}10^{16}$	$2.47{\cdot}10^{16}$	106710
297	5	$2.37{\cdot}10^{16}$	$2.47{\cdot}10^{16}$	118056
297	5	$2.64{\cdot}10^{16}$	$2.47{\cdot}10^{16}$	133935
297	2	$1.05{\cdot}10^{16}$	$9.87{\cdot}10^{15}$	47148
297	2	$9.48{\cdot}10^{15}$	$9.87{\cdot}10^{15}$	44071
297	2	$8.43{\cdot}10^{15}$	$9.87{\cdot}10^{15}$	37787
297	2	$7.38{\cdot}10^{15}$	$9.87{\cdot}10^{15}$	33301
297	2	$6.32{\cdot}10^{15}$	$9.87{\cdot}10^{15}$	28808
297	2	$5.27{\cdot}10^{15}$	$9.87{\cdot}10^{15}$	25455
297	2	$4.21{\cdot}10^{15}$	$9.87{\cdot}10^{15}$	21661

T / K	P / bar	[DMM] / cm^{-3}	[HNO_3] / cm^{-3}	k'_4 / s^{-1}
297	2	3.16·10^{15}	9.87·10^{15}	17555
297	2	2.11·10^{15}	9.86·10^{15}	13477
297	2	1.05·10^{15}	9.86·10^{15}	8862
297	2	0	9.86·10^{15}	4807
304	10	0	4.83·10^{16}	41792
304	10	5.16·10^{15}	4.83·10^{16}	68513
304	10	1.03·10^{16}	4.83·10^{16}	95008
304	10	1.55·10^{16}	4.82·10^{16}	114375
304	10	2.06·10^{16}	4.82·10^{16}	145130
304	10	2.57·10^{16}	4.82·10^{16}	169020
304	10	3.09·10^{16}	4.82·10^{16}	200727
304	10	3.60·10^{16}	4.82·10^{16}	234789
305	10	4.12·10^{16}	4.82·10^{16}	256058
305	10	4.63·10^{16}	4.82·10^{16}	273535
305	10	5.14·10^{16}	4.82·10^{16}	281184
305	5	2.57·10^{16}	2.41·10^{16}	117780
305	5	2.31·10^{16}	2.41·10^{16}	115586
305	5	2.06·10^{16}	2.41·10^{16}	107723
305	5	1.80·10^{16}	2.41·10^{16}	93765
305	5	1.54·10^{16}	2.40·10^{16}	83101
305	5	1.28·10^{16}	2.40·10^{16}	77132
305	5	1.03·10^{16}	2.40·10^{16}	62502
305	5	7.70·10^{15}	2.40·10^{16}	49684
305	5	5.13·10^{15}	2.40·10^{16}	39522
305	5	2.57·10^{15}	2.40·10^{16}	27613
305	5	0	2.40·10^{16}	17005
305	2	0	9.61·10^{15}	6175
305	2	1.03·10^{15}	9.61·10^{15}	10199
305	2	2.05·10^{15}	9.61·10^{15}	13665
305	2	3.08·10^{15}	9.61·10^{15}	18673
306	2	4.10·10^{15}	9.61·10^{15}	23279
306	2	5.13·10^{15}	9.60·10^{15}	26756
306	2	6.15·10^{15}	9.60·10^{15}	31317
306	2	7.18·10^{15}	9.60·10^{15}	35688
306	2	8.20·10^{15}	9.60·10^{15}	39126

T / K	P / bar	[DMM] / cm^{-3}	[HNO_3] / cm^{-3}	k_4' / s^{-1}
306	2	$9.23{\cdot}10^{15}$	$9.60{\cdot}10^{15}$	45785
306	2	$1.02{\cdot}10^{16}$	$9.60{\cdot}10^{15}$	47904
338	10	$4.63{\cdot}10^{16}$	$4.31{\cdot}10^{16}$	207235
338	10	$4.17{\cdot}10^{16}$	$4.31{\cdot}10^{16}$	199348
338	10	$3.71{\cdot}10^{16}$	$4.31{\cdot}10^{16}$	169712
338	10	$3.24{\cdot}10^{16}$	$4.31{\cdot}10^{16}$	146494
338	10	$2.78{\cdot}10^{16}$	$4.31{\cdot}10^{16}$	138914
338	10	$2.31{\cdot}10^{16}$	$4.30{\cdot}10^{16}$	117381
339	10	$1.85{\cdot}10^{16}$	$4.30{\cdot}10^{16}$	95713
339	10	$1.39{\cdot}10^{16}$	$4.30{\cdot}10^{16}$	83152
339	10	$9.25{\cdot}10^{15}$	$4.30{\cdot}10^{16}$	67440
339	10	$4.63{\cdot}10^{15}$	$4.30{\cdot}10^{16}$	41615
339	10	0	$4.30{\cdot}10^{16}$	25620
356	10	0	$4.09{\cdot}10^{16}$	18806
356	10	$4.40{\cdot}10^{15}$	$4.09{\cdot}10^{16}$	39287
356	10	$8.80{\cdot}10^{15}$	$4.09{\cdot}10^{16}$	56803
356	10	$1.32{\cdot}10^{16}$	$4.09{\cdot}10^{16}$	74077
356	10	$1.76{\cdot}10^{16}$	$4.09{\cdot}10^{16}$	86463
356	10	$2.20{\cdot}10^{16}$	$4.09{\cdot}10^{16}$	98224
356	10	$2.64{\cdot}10^{16}$	$4.09{\cdot}10^{16}$	130672
356	10	$3.08{\cdot}10^{16}$	$4.09{\cdot}10^{16}$	161552
356	10	$3.52{\cdot}10^{16}$	$4.10{\cdot}10^{16}$	166574
356	10	$3.97{\cdot}10^{16}$	$4.10{\cdot}10^{16}$	181904
355	10	$4.41{\cdot}10^{16}$	$4.10{\cdot}10^{16}$	199490
319	10	$4.91{\cdot}10^{16}$	$7.76{\cdot}10^{16}$	235331
319	10	$4.42{\cdot}10^{16}$	$7.76{\cdot}10^{16}$	286893
319	10	$3.93{\cdot}10^{16}$	$7.76{\cdot}10^{16}$	177621
319	10	$3.44{\cdot}10^{16}$	$7.76{\cdot}10^{16}$	192616
319	10	$2.94{\cdot}10^{16}$	$7.75{\cdot}10^{16}$	186022
319	10	$2.45{\cdot}10^{16}$	$7.76{\cdot}10^{16}$	141113
319	10	$1.96{\cdot}10^{16}$	$7.76{\cdot}10^{16}$	134145
319	10	$1.47{\cdot}10^{16}$	$7.76{\cdot}10^{16}$	100598
319	10	$9.81{\cdot}10^{15}$	$7.76{\cdot}10^{16}$	93175
319	10	$4.91{\cdot}10^{15}$	$7.75{\cdot}10^{16}$	60216
319	10	0	$7.75{\cdot}10^{16}$	31497

T / K	P / bar	[DMM] / cm^{-3}	[HNO_3] / cm^{-3}	k'_4 / s^{-1}
319	5	0	$3.87 \cdot 10^{16}$	14255
320	5	$2.45 \cdot 10^{15}$	$3.87 \cdot 10^{16}$	23382
320	5	$4.90 \cdot 10^{15}$	$3.87 \cdot 10^{16}$	37906
320	5	$7.35 \cdot 10^{15}$	$3.87 \cdot 10^{16}$	43522
320	5	$9.80 \cdot 10^{15}$	$3.87 \cdot 10^{16}$	54850
320	5	$1.23 \cdot 10^{16}$	$3.87 \cdot 10^{16}$	65700
320	5	$1.47 \cdot 10^{16}$	$3.87 \cdot 10^{16}$	73378
320	5	$1.72 \cdot 10^{16}$	$3.87 \cdot 10^{16}$	92391
320	5	$1.96 \cdot 10^{16}$	$3.87 \cdot 10^{16}$	100931
320	5	$2.21 \cdot 10^{16}$	$3.87 \cdot 10^{16}$	125386
320	5	$2.45 \cdot 10^{16}$	$3.87 \cdot 10^{16}$	107029
320	2	$9.80 \cdot 10^{15}$	$1.55 \cdot 10^{16}$	45998
320	2	$8.82 \cdot 10^{15}$	$1.55 \cdot 10^{16}$	39362
320	2	$7.84 \cdot 10^{15}$	$1.55 \cdot 10^{16}$	38132
320	2	$6.85 \cdot 10^{15}$	$1.55 \cdot 10^{16}$	34129
320	2	$5.88 \cdot 10^{15}$	$1.55 \cdot 10^{16}$	26492
320	2	$4.90 \cdot 10^{15}$	$1.55 \cdot 10^{16}$	22599
320	2	$3.92 \cdot 10^{15}$	$1.55 \cdot 10^{16}$	19750
320	2	$2.94 \cdot 10^{15}$	$1.55 \cdot 10^{16}$	17804
320	2	$1.96 \cdot 10^{15}$	$1.55 \cdot 10^{16}$	12203
320	2	$9.79 \cdot 10^{14}$	$1.55 \cdot 10^{16}$	7818
320	2	0	$1.55 \cdot 10^{16}$	5011
378	10	0	$6.54 \cdot 10^{16}$	24192
379	10	$4.13 \cdot 10^{15}$	$6.53 \cdot 10^{16}$	33552
380	10	$8.25 \cdot 10^{15}$	$6.52 \cdot 10^{16}$	56602
380	10	$1.24 \cdot 10^{16}$	$6.51 \cdot 10^{16}$	76882
381	10	$1.64 \cdot 10^{16}$	$6.50 \cdot 10^{16}$	110041
381	10	$2.05 \cdot 10^{16}$	$6.49 \cdot 10^{16}$	112624
382	10	$2.46 \cdot 10^{16}$	$6.48 \cdot 10^{16}$	124777
382	10	$2.87 \cdot 10^{16}$	$6.47 \cdot 10^{16}$	145815
382	10	$3.28 \cdot 10^{16}$	$6.48 \cdot 10^{16}$	192307
382	10	$3.69 \cdot 10^{16}$	$6.48 \cdot 10^{16}$	192447
382	10	$4.10 \cdot 10^{16}$	$6.47 \cdot 10^{16}$	183412
382	5	$2.05 \cdot 10^{16}$	$3.24 \cdot 10^{16}$	94561
381	5	$1.85 \cdot 10^{16}$	$3.25 \cdot 10^{16}$	92833

T / K	P / bar	[DMM] / cm^{-3}	[HNO_3] / cm^{-3}	k_4' / s^{-1}
382	5	$1.64 \cdot 10^{16}$	$3.24 \cdot 10^{16}$	77945
381	5	$1.44 \cdot 10^{16}$	$3.25 \cdot 10^{16}$	71401
381	5	$1.23 \cdot 10^{16}$	$3.25 \cdot 10^{16}$	59718
381	5	$1.03 \cdot 10^{16}$	$3.25 \cdot 10^{16}$	55615
381	5	$8.23 \cdot 10^{15}$	$3.25 \cdot 10^{16}$	42161
381	5	$6.17 \cdot 10^{15}$	$3.25 \cdot 10^{16}$	37077
381	5	$4.12 \cdot 10^{15}$	$3.25 \cdot 10^{16}$	22680
381	5	$2.06 \cdot 10^{15}$	$3.25 \cdot 10^{16}$	17264
380	5	0	$3.26 \cdot 10^{16}$	8534
380	2	0	$1.30 \cdot 10^{16}$	3434
380	2	$8.24 \cdot 10^{14}$	$1.30 \cdot 10^{16}$	6349
380	2	$1.65 \cdot 10^{15}$	$1.30 \cdot 10^{16}$	9943
380	2	$2.47 \cdot 10^{15}$	$1.30 \cdot 10^{16}$	11468
380	2	$3.30 \cdot 10^{15}$	$1.30 \cdot 10^{16}$	15677
380	2	$4.12 \cdot 10^{15}$	$1.30 \cdot 10^{16}$	21281
380	2	$4.95 \cdot 10^{15}$	$1.30 \cdot 10^{16}$	21941
380	2	$5.77 \cdot 10^{15}$	$1.30 \cdot 10^{16}$	25444
380	2	$6.60 \cdot 10^{15}$	$1.30 \cdot 10^{16}$	25472
380	2	$7.43 \cdot 10^{15}$	$1.30 \cdot 10^{16}$	34539
380	2	$8.25 \cdot 10^{15}$	$1.30 \cdot 10^{16}$	34512
357	2	$8.78 \cdot 10^{15}$	$1.39 \cdot 10^{16}$	33919
356	2	$7.91 \cdot 10^{15}$	$1.39 \cdot 10^{16}$	38783
357	2	$7.03 \cdot 10^{15}$	$1.39 \cdot 10^{16}$	29024
357	2	$6.15 \cdot 10^{15}$	$1.39 \cdot 10^{16}$	25509
357	2	$5.27 \cdot 10^{15}$	$1.39 \cdot 10^{16}$	23707
357	2	$4.39 \cdot 10^{15}$	$1.39 \cdot 10^{16}$	21571
356	2	$3.52 \cdot 10^{15}$	$1.39 \cdot 10^{16}$	17765
356	2	$2.64 \cdot 10^{15}$	$1.39 \cdot 10^{16}$	13908
356	2	$1.76 \cdot 10^{15}$	$1.39 \cdot 10^{16}$	11126
356	2	$8.80 \cdot 10^{14}$	$1.39 \cdot 10^{16}$	7111
356	2	0	$1.39 \cdot 10^{16}$	4043
356	5	0	$3.48 \cdot 10^{16}$	9426
356	5	$2.20 \cdot 10^{15}$	$3.48 \cdot 10^{16}$	15004
356	5	$4.40 \cdot 10^{15}$	$3.48 \cdot 10^{16}$	26528
356	5	$6.60 \cdot 10^{15}$	$3.48 \cdot 10^{16}$	36663

T / K	P / bar	[DMM] / cm^{-3}	[HNO_3] / cm^{-3}	k'_4 / s^{-1}
356	5	$8.80{\cdot}10^{15}$	$3.48{\cdot}10^{16}$	45057
356	5	$1.10{\cdot}10^{16}$	$3.48{\cdot}10^{16}$	53727
356	5	$1.32{\cdot}10^{16}$	$3.48{\cdot}10^{16}$	67669
356	5	$1.54{\cdot}10^{16}$	$3.48{\cdot}10^{16}$	84058
356	5	$1.76{\cdot}10^{16}$	$3.48{\cdot}10^{16}$	88726
356	5	$1.98{\cdot}10^{16}$	$3.48{\cdot}10^{16}$	94558
356	5	$2.20{\cdot}10^{16}$	$3.48{\cdot}10^{16}$	105599
406	10	0	$3.67{\cdot}10^{16}$	27666
406	10	$3.86{\cdot}10^{15}$	$3.66{\cdot}10^{16}$	41994
406	10	$7.71{\cdot}10^{15}$	$3.66{\cdot}10^{16}$	62685
407	10	$1.16{\cdot}10^{16}$	$3.66{\cdot}10^{16}$	65018
407	10	$1.54{\cdot}10^{16}$	$3.66{\cdot}10^{16}$	97286
407	10	$1.92{\cdot}10^{16}$	$3.66{\cdot}10^{16}$	121327
407	10	$2.31{\cdot}10^{16}$	$3.66{\cdot}10^{16}$	123332
407	10	$2.69{\cdot}10^{16}$	$3.66{\cdot}10^{16}$	120837
407	10	$3.08{\cdot}10^{16}$	$3.66{\cdot}10^{16}$	162232
407	10	$3.46{\cdot}10^{16}$	$3.66{\cdot}10^{16}$	179743
407	10	$3.85{\cdot}10^{16}$	$3.66{\cdot}10^{16}$	206310
407	5	$1.92{\cdot}10^{16}$	$1.83{\cdot}10^{16}$	93400
407	5	$1.73{\cdot}10^{16}$	$1.83{\cdot}10^{16}$	68678
407	5	$1.54{\cdot}10^{16}$	$1.83{\cdot}10^{16}$	69309
407	5	$1.35{\cdot}10^{16}$	$1.83{\cdot}10^{16}$	56399
407	5	$1.15{\cdot}10^{16}$	$1.83{\cdot}10^{16}$	70600
407	5	$9.62{\cdot}10^{15}$	$1.83{\cdot}10^{16}$	53186
407	5	$7.70{\cdot}10^{15}$	$1.83{\cdot}10^{16}$	42351
407	5	$5.77{\cdot}10^{15}$	$1.83{\cdot}10^{16}$	35663
407	5	$3.85{\cdot}10^{15}$	$1.83{\cdot}10^{16}$	27283
407	5	$1.92{\cdot}10^{15}$	$1.83{\cdot}10^{16}$	15857
407	5	0	$1.83{\cdot}10^{16}$	9637
407	2	0	$7.31{\cdot}10^{15}$	4020
407	2	$7.70{\cdot}10^{14}$	$7.31{\cdot}10^{15}$	6681
407	2	$1.54{\cdot}10^{15}$	$7.31{\cdot}10^{15}$	10187
407	2	$2.31{\cdot}10^{15}$	$7.32{\cdot}10^{15}$	12609
407	2	$3.08{\cdot}10^{15}$	$7.32{\cdot}10^{15}$	14537
407	2	$3.85{\cdot}10^{15}$	$7.32{\cdot}10^{15}$	18144

T / K	P / bar	[DMM] / cm^{-3}	[HNO_3] / cm^{-3}	k_4' / s^{-1}
407	2	$4.62{\cdot}10^{15}$	$7.32{\cdot}10^{15}$	20733
407	2	$5.39{\cdot}10^{15}$	$7.32{\cdot}10^{15}$	26617
407	2	$6.16{\cdot}10^{15}$	$7.32{\cdot}10^{15}$	24760
407	2	$6.93{\cdot}10^{15}$	$7.32{\cdot}10^{15}$	29622
407	2	$7.70{\cdot}10^{15}$	$7.32{\cdot}10^{15}$	29982
433	10	0	$3.44{\cdot}10^{16}$	22845
433	10	$3.62{\cdot}10^{15}$	$3.44{\cdot}10^{16}$	30848
433	10	$7.23{\cdot}10^{15}$	$3.43{\cdot}10^{16}$	49553
434	10	$1.08{\cdot}10^{16}$	$3.43{\cdot}10^{16}$	67532
434	10	$1.44{\cdot}10^{16}$	$3.43{\cdot}10^{16}$	79410
434	10	$1.80{\cdot}10^{16}$	$3.43{\cdot}10^{16}$	96433
435	10	$2.16{\cdot}10^{16}$	$3.42{\cdot}10^{16}$	130078
435	10	$2.52{\cdot}10^{16}$	$3.42{\cdot}10^{16}$	133097
435	10	$2.88{\cdot}10^{16}$	$3.42{\cdot}10^{16}$	139663
435	10	$3.24{\cdot}10^{16}$	$3.42{\cdot}10^{16}$	180146
435	10	$3.60{\cdot}10^{16}$	$3.42{\cdot}10^{16}$	183781
435	5	$1.80{\cdot}10^{16}$	$1.71{\cdot}10^{16}$	77328
435	5	$1.62{\cdot}10^{16}$	$1.71{\cdot}10^{16}$	73593
435	5	$1.44{\cdot}10^{16}$	$1.71{\cdot}10^{16}$	72426
435	5	$1.26{\cdot}10^{16}$	$1.71{\cdot}10^{16}$	64941
435	5	$1.08{\cdot}10^{16}$	$1.71{\cdot}10^{16}$	53083
435	5	$9.00{\cdot}10^{15}$	$1.71{\cdot}10^{16}$	46299
435	5	$7.21{\cdot}10^{15}$	$1.71{\cdot}10^{16}$	33216
435	5	$5.40{\cdot}10^{15}$	$1.71{\cdot}10^{16}$	29269
435	5	$3.60{\cdot}10^{15}$	$1.71{\cdot}10^{16}$	21749
435	5	$1.80{\cdot}10^{15}$	$1.71{\cdot}10^{16}$	13341
435	5	0	$1.71{\cdot}10^{16}$	8842
436	2	0	$6.83{\cdot}10^{15}$	3661
436	2	$7.19{\cdot}10^{14}$	$6.83{\cdot}10^{15}$	6693
436	2	$1.44{\cdot}10^{15}$	$6.83{\cdot}10^{15}$	8711
436	2	$2.16{\cdot}10^{15}$	$6.83{\cdot}10^{15}$	11102
436	2	$2.88{\cdot}10^{15}$	$6.83{\cdot}10^{15}$	12952
436	2	$3.60{\cdot}10^{15}$	$6.83{\cdot}10^{15}$	17687
436	2	$4.31{\cdot}10^{15}$	$6.83{\cdot}10^{15}$	18901
436	2	$5.03{\cdot}10^{15}$	$6.83{\cdot}10^{15}$	25271

T / K	P / bar	[DMM] / cm^{-3}	[HNO_3] / cm^{-3}	k'_4 / s^{-1}
436	2	$5.75 \cdot 10^{15}$	$6.83 \cdot 10^{15}$	23738
436	2	$6.47 \cdot 10^{15}$	$6.83 \cdot 10^{15}$	24433
436	2	$7.19 \cdot 10^{15}$	$6.83 \cdot 10^{15}$	28443
457	5	$1.71 \cdot 10^{16}$	$1.63 \cdot 10^{16}$	77672
458	5	$1.54 \cdot 10^{16}$	$1.63 \cdot 10^{16}$	68000
458	5	$1.37 \cdot 10^{16}$	$1.62 \cdot 10^{16}$	69296
459	5	$1.20 \cdot 10^{16}$	$1.62 \cdot 10^{16}$	58483
459	5	$1.02 \cdot 10^{16}$	$1.62 \cdot 10^{16}$	51327
459	5	$8.53 \cdot 10^{15}$	$1.62 \cdot 10^{16}$	38916
459	5	$6.82 \cdot 10^{15}$	$1.62 \cdot 10^{16}$	33527
459	5	$5.12 \cdot 10^{15}$	$1.62 \cdot 10^{16}$	33307
459	5	$3.41 \cdot 10^{15}$	$1.62 \cdot 10^{16}$	20351
459	5	$1.71 \cdot 10^{15}$	$1.62 \cdot 10^{16}$	15090
459	5	0	$1.62 \cdot 10^{16}$	8199
459	2	0	$6.49 \cdot 10^{15}$	3695
459	2	$6.83 \cdot 10^{14}$	$6.49 \cdot 10^{15}$	6306
458	2	$1.37 \cdot 10^{15}$	$6.49 \cdot 10^{15}$	9128
458	2	$2.05 \cdot 10^{15}$	$6.49 \cdot 10^{15}$	12318
458	2	$2.74 \cdot 10^{15}$	$6.50 \cdot 10^{15}$	13020
458	2	$3.42 \cdot 10^{15}$	$6.50 \cdot 10^{15}$	16995
458	2	$4.11 \cdot 10^{15}$	$6.50 \cdot 10^{15}$	21709
458	2	$4.79 \cdot 10^{15}$	$6.50 \cdot 10^{15}$	21887
458	2	$5.48 \cdot 10^{15}$	$6.50 \cdot 10^{15}$	20955
457	2	$6.16 \cdot 10^{15}$	$6.51 \cdot 10^{15}$	25018
457	2	$6.85 \cdot 10^{15}$	$6.51 \cdot 10^{15}$	31857
338	2	0	$8.80 \cdot 10^{15}$	7017
340	2	$1.04 \cdot 10^{15}$	$8.77 \cdot 10^{15}$	9526
339	2	$2.08 \cdot 10^{15}$	$8.78 \cdot 10^{15}$	13888
339	2	$3.12 \cdot 10^{15}$	$8.77 \cdot 10^{15}$	15507
339	2	$4.17 \cdot 10^{15}$	$8.78 \cdot 10^{15}$	21793
339	2	$5.21 \cdot 10^{15}$	$8.78 \cdot 10^{15}$	24633
339	2	$6.25 \cdot 10^{15}$	$8.78 \cdot 10^{15}$	27262
339	2	$7.29 \cdot 10^{15}$	$8.78 \cdot 10^{15}$	31890
339	2	$8.33 \cdot 10^{15}$	$8.78 \cdot 10^{15}$	35525
339	2	$9.37 \cdot 10^{15}$	$8.77 \cdot 10^{15}$	42615

T / K	P / bar	[DMM] / cm^{-3}	[HNO_3] / cm^{-3}	k'_4 / s^{-1}
339	2	$1.04{\cdot}10^{16}$	$8.77{\cdot}10^{15}$	38517
339	5	$2.60{\cdot}10^{16}$	$2.19{\cdot}10^{16}$	118572
339	5	$2.34{\cdot}10^{16}$	$2.19{\cdot}10^{16}$	106010
339	5	$2.08{\cdot}10^{16}$	$2.19{\cdot}10^{16}$	99393
339	5	$1.82{\cdot}10^{16}$	$2.19{\cdot}10^{16}$	96575
339	5	$1.56{\cdot}10^{16}$	$2.19{\cdot}10^{16}$	82174
339	5	$1.30{\cdot}10^{16}$	$2.19{\cdot}10^{16}$	70467
339	5	$1.04{\cdot}10^{16}$	$2.19{\cdot}10^{16}$	61573
339	5	$7.80{\cdot}10^{15}$	$2.19{\cdot}10^{16}$	51157
339	5	$5.20{\cdot}10^{15}$	$2.19{\cdot}10^{16}$	40694
339	5	$2.60{\cdot}10^{15}$	$2.19{\cdot}10^{16}$	29862
339	5	0	$2.19{\cdot}10^{16}$	20625
490	5	0	$1.52{\cdot}10^{16}$	8850
491	5	$1.80{\cdot}10^{15}$	$1.52{\cdot}10^{16}$	17486
490	5	$3.60{\cdot}10^{15}$	$1.52{\cdot}10^{16}$	23029
490	5	$5.41{\cdot}10^{15}$	$1.52{\cdot}10^{16}$	28415
489	5	$7.21{\cdot}10^{15}$	$1.52{\cdot}10^{16}$	34560
489	5	$9.02{\cdot}10^{15}$	$1.52{\cdot}10^{16}$	43956
490	5	$1.08{\cdot}10^{16}$	$1.52{\cdot}10^{16}$	51641
490	5	$1.26{\cdot}10^{16}$	$1.52{\cdot}10^{16}$	63993
490	5	$1.44{\cdot}10^{16}$	$1.52{\cdot}10^{16}$	63831
490	5	$1.62{\cdot}10^{16}$	$1.52{\cdot}10^{16}$	70566
490	5	$1.80{\cdot}10^{16}$	$1.52{\cdot}10^{16}$	85539
490	2	$7.20{\cdot}10^{15}$	$6.07{\cdot}10^{15}$	33130
490	2	$6.49{\cdot}10^{15}$	$6.08{\cdot}10^{15}$	28034
490	2	$5.76{\cdot}10^{15}$	$6.08{\cdot}10^{15}$	25482
490	2	$5.04{\cdot}10^{15}$	$6.07{\cdot}10^{15}$	22914
490	2	$4.32{\cdot}10^{15}$	$6.07{\cdot}10^{15}$	20823
490	2	$3.60{\cdot}10^{15}$	$6.07{\cdot}10^{15}$	18274
490	2	$2.88{\cdot}10^{15}$	$6.07{\cdot}10^{15}$	15613
490	2	$2.16{\cdot}10^{15}$	$6.07{\cdot}10^{15}$	13616
490	2	$1.44{\cdot}10^{15}$	$6.07{\cdot}10^{15}$	10398
490	2	$7.21{\cdot}10^{14}$	$6.07{\cdot}10^{15}$	7733
490	2	0	$6.08{\cdot}10^{15}$	3340
528	5	0	$1.41{\cdot}10^{16}$	11214

T / K	P / bar	[DMM] / cm^{-3}	[HNO_3] / cm^{-3}	k'_4 / s^{-1}
528	5	$1.67 \cdot 10^{15}$	$1.41 \cdot 10^{16}$	21197
529	5	$3.34 \cdot 10^{15}$	$1.41 \cdot 10^{16}$	26309
529	5	$5.01 \cdot 10^{15}$	$1.41 \cdot 10^{16}$	32635
529	5	$6.68 \cdot 10^{15}$	$1.41 \cdot 10^{16}$	34596
529	5	$8.34 \cdot 10^{15}$	$1.41 \cdot 10^{16}$	43414
529	5	$1.00 \cdot 10^{16}$	$1.41 \cdot 10^{16}$	49378
529	5	$1.17 \cdot 10^{16}$	$1.41 \cdot 10^{16}$	56979
529	5	$1.33 \cdot 10^{16}$	$1.41 \cdot 10^{16}$	61762
529	5	$1.50 \cdot 10^{16}$	$1.41 \cdot 10^{16}$	68132
530	5	$1.67 \cdot 10^{16}$	$1.40 \cdot 10^{16}$	71670
529	2	$6.67 \cdot 10^{15}$	$5.62 \cdot 10^{15}$	26937
529	2	$6.00 \cdot 10^{15}$	$5.62 \cdot 10^{15}$	25070
529	2	$5.33 \cdot 10^{15}$	$5.62 \cdot 10^{15}$	23143
529	2	$4.67 \cdot 10^{15}$	$5.62 \cdot 10^{15}$	20100
529	2	$4.00 \cdot 10^{15}$	$5.62 \cdot 10^{15}$	21545
529	2	$3.34 \cdot 10^{15}$	$5.62 \cdot 10^{15}$	19077
529	2	$2.67 \cdot 10^{15}$	$5.62 \cdot 10^{15}$	16037
529	2	$2.00 \cdot 10^{15}$	$5.62 \cdot 10^{15}$	12632
529	2	$1.33 \cdot 10^{15}$	$5.62 \cdot 10^{15}$	11307
528	2	$6.68 \cdot 10^{14}$	$5.64 \cdot 10^{15}$	8054
528	2	0	$5.63 \cdot 10^{15}$	4597
569	2	0	$5.23 \cdot 10^{15}$	5369
569	2	$6.21 \cdot 10^{14}$	$5.23 \cdot 10^{15}$	8997
568	2	$1.24 \cdot 10^{15}$	$5.24 \cdot 10^{15}$	11031
568	2	$1.86 \cdot 10^{15}$	$5.24 \cdot 10^{15}$	12866
568	2	$2.49 \cdot 10^{15}$	$5.24 \cdot 10^{15}$	16191
567	2	$3.11 \cdot 10^{15}$	$5.25 \cdot 10^{15}$	15267
567	2	$3.74 \cdot 10^{15}$	$5.25 \cdot 10^{15}$	19965
567	2	$4.36 \cdot 10^{15}$	$5.25 \cdot 10^{15}$	19622
567	2	$4.98 \cdot 10^{15}$	$5.25 \cdot 10^{15}$	19590
567	2	$5.60 \cdot 10^{15}$	$5.25 \cdot 10^{15}$	22672
567	2	$6.23 \cdot 10^{15}$	$5.25 \cdot 10^{15}$	25134
459	10	0	$3.03 \cdot 10^{16}$	17289
459	10	$3.84 \cdot 10^{15}$	$3.03 \cdot 10^{16}$	42682
459	10	$7.68 \cdot 10^{15}$	$3.03 \cdot 10^{16}$	55824

T / K	P / bar	[DMM] / cm^{-3}	[HNO_3] / cm^{-3}	k_4' / s^{-1}
459	10	$1.15 \cdot 10^{16}$	$3.03 \cdot 10^{16}$	65266
459	10	$1.54 \cdot 10^{16}$	$3.03 \cdot 10^{16}$	75761
459	10	$1.92 \cdot 10^{16}$	$3.02 \cdot 10^{16}$	87314
460	10	$2.30 \cdot 10^{16}$	$3.02 \cdot 10^{16}$	103290
460	10	$2.69 \cdot 10^{16}$	$3.02 \cdot 10^{16}$	115916
459	10	$3.07 \cdot 10^{16}$	$3.03 \cdot 10^{16}$	129281
458	10	$3.47 \cdot 10^{16}$	$3.03 \cdot 10^{16}$	134461
458	10	$3.85 \cdot 10^{16}$	$3.03 \cdot 10^{16}$	164175
489	10	$3.61 \cdot 10^{16}$	$2.84 \cdot 10^{16}$	152256
489	10	$3.25 \cdot 10^{16}$	$2.84 \cdot 10^{16}$	165311
489	10	$2.88 \cdot 10^{16}$	$2.84 \cdot 10^{16}$	153947
490	10	$2.52 \cdot 10^{16}$	$2.84 \cdot 10^{16}$	128007
490	10	$2.16 \cdot 10^{16}$	$2.83 \cdot 10^{16}$	117738
491	10	$1.80 \cdot 10^{16}$	$2.83 \cdot 10^{16}$	103054
491	10	$1.44 \cdot 10^{16}$	$2.83 \cdot 10^{16}$	75466
491	10	$1.08 \cdot 10^{16}$	$2.83 \cdot 10^{16}$	65940
491	10	$7.19 \cdot 10^{15}$	$2.83 \cdot 10^{16}$	53654
491	10	$3.59 \cdot 10^{15}$	$2.83 \cdot 10^{16}$	40437
490	10	0	$2.84 \cdot 10^{16}$	23015
530	10	0	$2.62 \cdot 10^{16}$	39233
530	10	$3.33 \cdot 10^{15}$	$2.62 \cdot 10^{16}$	67374
530	10	$6.66 \cdot 10^{15}$	$2.62 \cdot 10^{16}$	88722
530	10	$1.00 \cdot 10^{16}$	$2.62 \cdot 10^{16}$	101260
529	10	$1.33 \cdot 10^{16}$	$2.63 \cdot 10^{16}$	102194
529	10	$1.67 \cdot 10^{16}$	$2.63 \cdot 10^{16}$	122714
529	10	$2.00 \cdot 10^{16}$	$2.63 \cdot 10^{16}$	143005
529	10	$2.33 \cdot 10^{16}$	$2.63 \cdot 10^{16}$	142456
529	10	$2.67 \cdot 10^{16}$	$2.63 \cdot 10^{16}$	163766
529	10	$3.00 \cdot 10^{16}$	$2.63 \cdot 10^{16}$	174368
529	10	$3.33 \cdot 10^{16}$	$2.63 \cdot 10^{16}$	201693
568	10	$3.11 \cdot 10^{16}$	$2.45 \cdot 10^{16}$	179726
568	10	$2.80 \cdot 10^{16}$	$2.45 \cdot 10^{16}$	169242
568	10	$2.49 \cdot 10^{16}$	$2.45 \cdot 10^{16}$	146122
568	10	$2.17 \cdot 10^{16}$	$2.45 \cdot 10^{16}$	140066
568	10	$1.86 \cdot 10^{16}$	$2.45 \cdot 10^{16}$	132296

T / K	P / bar	[DMM] / cm^{-3}	[HNO_3] / cm^{-3}	k'_4 / s^{-1}
569	10	$1.55 \cdot 10^{16}$	$2.44 \cdot 10^{16}$	127722
569	10	$1.24 \cdot 10^{16}$	$2.44 \cdot 10^{16}$	117014
570	10	$9.29 \cdot 10^{15}$	$2.44 \cdot 10^{16}$	100673
570	10	$6.19 \cdot 10^{15}$	$2.44 \cdot 10^{16}$	101567
571	10	$3.09 \cdot 10^{15}$	$2.44 \cdot 10^{16}$	77672
570	10	0	$2.44 \cdot 10^{16}$	48086
571	5	0	$1.22 \cdot 10^{16}$	15990
571	5	$1.55 \cdot 10^{15}$	$1.22 \cdot 10^{16}$	22511
571	5	$3.09 \cdot 10^{15}$	$1.22 \cdot 10^{16}$	31003
571	5	$4.64 \cdot 10^{15}$	$1.22 \cdot 10^{16}$	30776
570	5	$6.20 \cdot 10^{15}$	$1.22 \cdot 10^{16}$	39342
570	5	$7.74 \cdot 10^{15}$	$1.22 \cdot 10^{16}$	46114
570	5	$9.29 \cdot 10^{15}$	$1.22 \cdot 10^{16}$	52035
570	5	$1.08 \cdot 10^{16}$	$1.22 \cdot 10^{16}$	57820
570	5	$1.24 \cdot 10^{16}$	$1.22 \cdot 10^{16}$	57531
570	5	$1.39 \cdot 10^{16}$	$1.22 \cdot 10^{16}$	64759
570	5	$1.55 \cdot 10^{16}$	$1.22 \cdot 10^{16}$	71761

Table A.8: Overview of the experimental conditions in the measurements of the gas flow dependence of k_4' in the system DMM + OH.

T / K	P / bar	f / slm	[DMM] / cm^{-3}	[HNO_3] / cm^{-3}	k_4' / s^{-1}
297	10	0.7	$2.64 \cdot 10^{16}$	$4.94 \cdot 10^{16}$	151481
297	10	1.4	$2.64 \cdot 10^{16}$	$4.94 \cdot 10^{16}$	165616
297	10	2.1	$2.64 \cdot 10^{16}$	$4.94 \cdot 10^{16}$	167201
297	10	2.8	$2.64 \cdot 10^{16}$	$4.94 \cdot 10^{16}$	169124
297	10	3.5	$2.64 \cdot 10^{16}$	$4.94 \cdot 10^{16}$	163288
297	10	4.9	$2.64 \cdot 10^{16}$	$4.94 \cdot 10^{16}$	163355
297	10	6.3	$2.64 \cdot 10^{16}$	$4.94 \cdot 10^{16}$	160750
297	10	7.7	$2.64 \cdot 10^{16}$	$4.94 \cdot 10^{16}$	157916
297	10	9.1	$2.64 \cdot 10^{16}$	$4.94 \cdot 10^{16}$	150654
565	2	0.7	$3.13 \cdot 10^{15}$	$4.92 \cdot 10^{15}$	15370
565	2	1.4	$3.12 \cdot 10^{15}$	$4.92 \cdot 10^{15}$	15350
566	2	2.1	$3.12 \cdot 10^{15}$	$4.91 \cdot 10^{15}$	17655
566	2	2.8	$3.12 \cdot 10^{15}$	$4.91 \cdot 10^{15}$	18042
565	2	3.5	$3.13 \cdot 10^{15}$	$4.92 \cdot 10^{15}$	16927
565	2	4.9	$3.12 \cdot 10^{15}$	$4.92 \cdot 10^{15}$	18598
566	2	6.3	$3.12 \cdot 10^{15}$	$4.91 \cdot 10^{15}$	17757
568	2	7.7	$3.11 \cdot 10^{15}$	$4.89 \cdot 10^{15}$	17604
568	2	9.1	$3.11 \cdot 10^{15}$	$4.89 \cdot 10^{15}$	21553

Table A.9: Overview of the experimental conditions in the measurements of the repetition rate dependence of k_4' in the system DMM + OH.

T / K	P / bar	ν / Hz	[DMM] / cm^{-3}	[HNO_3] / cm^{-3}	k_4' / s^{-1}
294	10	10	$3.00 \cdot 10^{16}$	$4.82 \cdot 10^{16}$	220616
294	10	5	$3.00 \cdot 10^{16}$	$4.82 \cdot 10^{16}$	216331
294	10	1	$3.00 \cdot 10^{16}$	$4.82 \cdot 10^{16}$	208715
565	2	10	$3.13 \cdot 10^{15}$	$4.92 \cdot 10^{15}$	16927
567	2	5	$3.11 \cdot 10^{15}$	$4.90 \cdot 10^{15}$	15444
566	2	1	$3.12 \cdot 10^{15}$	$4.91 \cdot 10^{15}$	16873

Best fits of a simple Arrhenius expression to the bimolecular rate coefficients k_4 at different pressures:

$$k_4(2\ \text{bar}, T) = 2.46 \cdot 10^{-12} \exp\left(\frac{153\ \text{K}}{T}\right)\ \text{cm}^3\ \text{s}^{-1} \tag{A.1}$$

$$k_4(5\ \text{bar}, T) = 3.12 \cdot 10^{-12} \exp\left(\frac{105\ \text{K}}{T}\right)\ \text{cm}^3\ \text{s}^{-1} \tag{A.2}$$

$$k_4(10\ \text{bar}, T) = 3.09 \cdot 10^{-12} \exp\left(\frac{126\ \text{K}}{T}\right)\ \text{cm}^3\ \text{s}^{-1}\,. \tag{A.3}$$

Bibliography

[1] T. E. Graedel and P. J. Crutzen, *Atmospheric Change – An Earth System Perspective*, **1993**, W. H. Freeman and Companny: New York.

[2] F. Joos, *Technische Verbrennung – Verbrennungstechnik, Verbrennungsmodellierung, Emissionen*, **2006**, Springer Verlag: Berlin.

[3] P. Muller, *Pure Appl. Chem.* **1994**, *66*, 1077-1184.

[4] M. Binnewies, *Allgemeine und Anorganische Chemie*, **2016**, 3rd Ed., Springer Spektrum: Berlin.

[5] D. R. Lide, *CRC Handbook of Chemistry and Physics*, **1993**, 74th Ed., CRC Press: Boca Raton.

[6] G. P. Merker and R. Teichmann, *Grundlagen Verbrennungsmotoren*, **2014**, 7th Ed., Springer Vieweg: Wiesbaden.

[7] J. H. Seinfeld and S. N. Pandis, *Atmospheric Chemistry and Physics: From Air Pollution to Climate Change*, **2016**, Wiley: Hoboken.

[8] D. H. Ehhalt, *Phys. Chem. Chem. Phys.* **1999**, *1*, 5401-5408.

[9] R. G. Prinn, *Annu. Rev. Environ. Resour.* **2003**, *28*, 29-57.

[10] S. S. Brown, R. K. Talukdar and A. R. Ravishankara, *J. Phys. Chem. A* **1999**, *103*, 3031-3037.

[11] S. A. Carr, M. T. Baeza-Romero, M. A. Blitz, B. J. S. Price and P. W. Seakins, *Int. J. Chem. Kinet.* **2008**, *40*, 504-514.

[12] B. Derstroff, *Modellierung des Zündverhaltens diethyletherhaltiger Mischungen*, Internship Report, Karlsruher Institut für Technologie, **2012**.

[13] J. Hetzler, *Untersuchungen zur Kinetik der Reaktion OH mit Diethylether bei hohen Drücken*, Internship Report, Karlsruher Institut für Technologie, **2011**.

[14] C. Hüllemann, *Untersuchungen zur Kinetik der Reaktionen von Diethylether und Dimethylether mit Hydroxylradikalen*, Diploma thesis, Karlsruher Institut für Technologie, **2012**.

[15] K. Kohse-Höinghaus, P. Owald, T. A. Cool, T. Kasper, N. Hansen, F. Qi, C. K. Westbrook and P. R. Westmoreland, *Angew. Chem. Int. Ed.* **2010**, *49*, 3572-3597.

[16] A. Arteconi, A. Mazzarini and G. Di Nicola, *Water Air Soil Pollut.* **2011**, *221*, 405-423.

[17] National Research Council (U.S.), *Prudent Practices in the Laboratory: Handling and Disposal of Chemicals*, **1995**, National Academy Press: Washington.

[18] J. Kiecherer, *Experimentelle und theoretische Untersuchungen zur Kinetik der Pyrolyse und Oxidation von Diethylether*, Ph.D. thesis, Karlsruher Institut für Technologie, **2015**.

[19] W. Demtröder, *Laserspektroskopie 1*, **2011**, 6th Ed., Springer-Verlag: Heidelberg.

[20] W. Demtröder, *Laserspektroskopie 2*, **2013**, 6th Ed., Springer-Verlag: Heidelberg.

[21] M. Kasha, *Discuss. Faraday Soc.* **1950**, *9*, 14-19.

[22] J. W. Daily, *Appl. Opt.* **1977**, *16*, 568-571.

[23] A. C. Eckbreth, *Laser Diagnostics for Combustion Temperature and Species*, **1988**, Abacus Press: Kent.

[24] P. Bouguer, *Essai d'optique, Sur la gradation de la lumière*, **1729**, C. Jombert: Paris.

[25] J. H. Lambert, *Photometria, sive de mensura et gradibus luminis, colorum et umbrae*, **1760**, Sumptibus Vidae Eberhardi Klett: Straßburg.

[26] A. Beer, *Ann. Phys.* **1852**, *86*, 78-88.

[27] J. B. Burkholder, S. P. Sander, J. Abbatt, J. R. Barker, R. E. Huie, C. E. Kolb, M. J. Kurylo, V. L. Orkin, D. M. Wilmouth and P. H. Wine, *Chemical Kinetics and Photochemical Data for Use in Atmospheric Studies, Evaluation No. 18*, **2015**, JPL Publication 15-10, Jet Propulsion Laboratory, http://jpldataeval.jpl.nasa.gov.

[28] R. Forster, *Rekombinationsreaktionen des Hydroxyl-Radikals bei hohen Drücken: Eine Anwendung der Laserspektroskopie*, Ph.D. thesis, Universität Göttingen, **1991**.

[29] S. Arrhenius, *Z. Phys. Chem.* **1889**, *4*, 226-248.

[30] P. L. Houston, *Chemical Kinetics and Reaction Dynamics*, **2001**, Dover publications, Inc.: Mineola.

[31] D. M. Kooij, *Z. Phys. Chem.* **1893**, *12*, 155-161.

[32] I. W. M. Smith, *Chem. Soc. Rev.* **2008**, *37*, 812-826.

[33] F. A. Lindemann, I. Langmuir, N. R. Dhar, J. Perrin and W. C. M. Lewis, *Trans. Faraday Soc.* **1922**, *17*, 598-606.

[34] J. Troe, *J. Phys. Chem.* **1979**, *83*, 114-126.

[35] H. Eyring, *J. Chem. Phys.* **1935**, *3*, 107-115.

[36] M. G. Evans and M. Polanyi, *Trans. Faraday Soc.* **1935**, *31*, 875-894.

[37] O. K. Rice and H. C. Ramsperger, *J. Am. Chem. Soc.* **1927**, *49*, 1617-1629.

[38] L. S. Kassel, *J. Phys. Chem.* **1928**, *32*, 1065-1079.

[39] R. A. Marcus and O. K. Rice, *J. Phys. Chem.* **1951**, *55*, 894-908.

[40] M. Quack and J. Troe, *Ber. Bunsenges. Phys. Chem.* **1974**, *78*, 240-252.

[41] M. Hoare, *J. Chem. Phys.* **1963**, *38*, 1630-1635.

[42] G. H. Kohlmaier and B. S. Rabinovitch, *J. Chem. Phys.* **1963**, *38*, 1692-1708.

[43] R. G. Gilbert and S. C. Smith, *Theory of Unimolecular and Recombination Reactions*, **1990**, Blackwell Scientific Publishing: Oxford.

[44] T. Baer and W. L. Hase, *Unimolecular Reaction Dynamics*, **1996**, Oxford University Press: Oxford.

[45] J. I. Steinfeld, J. S. Francisco and W. L. Hase, *Chemical Kinetics and Dynamics*, **1999**, 2nd Ed., Prentice Hall: New Jersey.

[46] H. W. Schranz and S. Nordholm, *Chem. Phys.* **1984**, *87*, 163-177.

[47] J. Troe, *J. Chem. Soc. Faraday Trans.* **1994**, *90*, 2303-2317.

[48] M. Olzmann, in: F. Battin-Leclerc, J. M. Simmie and E. Blurock (Eds.), *Cleaner Combustion – Developing Detailed Chemical Kinetic Models*, **2013**, Springer: London, 549-576.

[49] F. Jensen, *Introduction to Computational Chemistry*, **2007**, 2nd Ed., John Wiley & Sons: Chichester.

[50] E. Schrödinger, *Ann. Phys.* **1926**, *384*, 361-376.

[51] M. Born and R. Oppenheimer, *Ann. Phys.* **1927**, *389*, 457-484.

[52] F. Fock, *Z. Phys.* **1930**, *61*, 126-148.

[53] J. Čížek, *J. Chem. Phys.* **1966**, *45*, 4256-4266.

[54] P. Hohenberg and W. Kohn, *Phys. Rev.* **1964**, *136*, B864-B871.

[55] W. Kohn and L. J. Sham, *Phys. Rev.* **1965**, *140*, A1133-A1138.

[56] W. Koch and M. C. Holthausen, *A Chemist's Guide to Density Functional Theory*, **2001**, 2nd Ed., Wiley-VCH Verlag GmbH: Weinheim.

[57] A. D. Becke, *J. Chem. Phys.* **1993**, *98*, 1372-1377.

[58] A. D. Becke, *J. Chem. Phys.* **1993**, *98*, 5648-5652.

[59] R. A. Friesner, *Proc. Natl. Acad. Sci. USA* **2005**, *102*, 6648-6653.

[60] S. Grimme, *Wiley Interdiscip. Rev. Comput. Mol. Sci.* **2011**, *1*, 211-228.

[61] Q. Zhang, R. Bell and T. N. Truong, *J. Phys. Chem.* **1995**, *99*, 592-599.

[62] B. S. Jursic, *Chem. Phys. Lett.* **1997**, *264*, 113-119.

[63] B. J. Lynch, P. L. Fast, M. Harris and D. G. Truhlar, *J. Phys. Chem. A* **2000**, *104*, 4811-4815.

[64] F. Striebel, *Experimentelle Untersuchungen zur Kinetik von Elementarreaktionen unter Hochdruckbedingungen – Ein Beitrag zum Verständnis unimolekularer Reaktionen*, Ph.D. thesis, Universität Karlsruhe (TH), **2000**.

[65] C. Kappler, *Untersuchungen komplexbildender bimolekularer Reaktionen in der Gasphase mit laserspektroskopischen Methoden und statistischer Reaktionstheorie*, Ph.D. thesis, Karlsruher Institut für Technologie, **2010**.

[66] J. Sommerer, *Kinetische Untersuchungen von Reaktionen kurzlebiger Intermediate im Zündfunken und bei der Verbrennung*, Ph.D. thesis, Karlsruher Institut für Technologie, **2011**.

[67] H. Hippler, N. Krasteva and F. Striebel, *Phys. Chem. Chem. Phys.* **2004**, *6*, 3383-3388.

[68] L. N. Krasnoperov, *Phys. Chem. Chem. Phys.* **2005**, *7*, 2074-2076.

[69] H. Hippler, N. Krasteva and F. Striebel, *Phys. Chem. Chem. Phys.* **2005**, *7*, 2077-2079.

[70] A. Stern, J. T. Multhaupt and W. B. Kay, *Chem. Rev.* **1960**, *60*, 185-207.

[71] R. A. Stachnik, L. T. Molina and M. J. Molina, *J . Phys. Chem.* **1986**, *90*, 2777-2780.

[72] A. J. C. Bunkan, J. Hetzler, T. Mikoviny, A. Wisthaler, C. J. Nielsen and M. Olzmann, *Phys. Chem. Chem. Phys.* **2015**, *17*, 7046-7059.

[73] C. Bänsch, P. Hibomvschi and M. Olzmann, *Hochdrucksättiger für korrosive und hochreine Flüssigkeiten* **2017**, patent pending, appl. no. 102017123985.5.

[74] J. Hetzler, *Untersuchungen zur Kinetik bimolekularer Reaktionen und stoßinduzierter Relaxationsprozesse mittels zeitaufgelöster Laserspektroskopie*, Ph.D. thesis, Karlsruher Institut für Technologie, **2016**.

[75] D. Fulle, *Untersuchungen zur Kinetik von Reaktionen des OH- und des CH-Radikals unter Hochdruckbedingungen*, Ph.D. thesis, Universität Göttingen, **1996**.

[76] G. L. Zügner, *private communication*, **2016**.

[77] I. Acalovschi, *Zum Einfluß von CO_2 auf die Geschwindigkeitskonstanten und Gleichgewichte chemischer Reaktionen im Cl/O_2-System*, Ph.D. thesis, Universität Karlsruhe, **2000**.

[78] J. R. Huber, *Chem. Phys. Chem.* **2004**, *5*, 1663-1669.

[79] A. A. Turnipseed, G. L. Vaghjiani, J. E. Thompson and A. R. Ravishankara, *J. Chem. Phys.* **1992**, *96*, 5887-5895.

[80] V. Riffault, T. Gierczak, J. B. Burkholder and A. R. Ravishankara, *Phys. Chem. Chem. Phys.* **2006**, *8*, 1079-1085.

[81] H. Hippler, S. Nasterlack and F. Striebel, *Phys. Chem. Chem. Phys.* **2002**, *4*, 2959-2964.

[82] C. Kappler, J. Zádor, O. Welz, R. X. Fernandes, M. Olzmann and C. A. Taatjes, *Z. Phys. Chem.* **2011**, *225*, 1271-1291.

[83] J. B. Burkholder, R. K. Talukdar, A. R. Ravishankara and S. Solomon, *J. Geophys. Res. D* **1993**, *98*, 22937-22948.

[84] D. C. McCabe, S. S. Brown, M. K. Gilles, R. K. Talukdar, I. W. M. Smith and A. R. Ravishankara, *J. Phys. Chem. A* **2003**, *107*, 7762-7769.

[85] R. Atkinson, *Chem. Rev.* **1985**, *85*, 69-201.

[86] K. Kohse-Höinghaus and J. B. Jeffries, *Applied Combustion Diagnostics*, **2002**, Taylor & Francis: New York.

[87] S. Nasterlack, *Bestimmung von Ausbeuten bei komplexbildenden bimolekularen Reaktionen mittels laserinduzierter Fluoreszenz*, Ph.D. thesis, Universität Karlsruhe (TH), **2004**.

[88] W. J. Moore and D. O. Hummel, *Physikalische Chemie*, **1986**, 4th Ed., de Gruyter: Berlin.

[89] H. K. Chung and A. Dalgarno, *Phys. Rev. A* **2002**, *66*, 012712-1-012712-3.

[90] M. J. Pilling and P. W. Seakins, *Reaction Kinetics*, **1995**, Oxford Scientific Publications: Oxford.

[91] R. G. W. Norrish and G. Porter, *Nature* **1949**, *164*, 658.

[92] J. L. Kinsey, *Ann. Rev. Phys. Chem.* **1977**, *28*, 349-372.

[93] S. Gligorovski, R. Strekowski, S. Barbati and D. Vione, *Chem. Rev.* **2015**, *115*, 13051-13092.

[94] R. Forster, M. Frost, D. Fulle, H. F. Hamann, H. Hippler, A. Schlepegrell and J. Troe, *J. Chem. Phys.* **1995**, *103*, 2949-2958.

[95] M. Arif, B. Dellinger and P. H. Taylor, *J. Phys. Chem. A* **1997**, *101*, 2436-2441.

[96] S. A. Carr, T. J. Still, M. A. Blitz, A. J. Eskola, M. J. Pilling, P. W. Seakins, R. J. Shannon, B. Wang and S. H. Robertson, *J. Phys. Chem. A* **2013**, *117*, 11142-11154.

[97] D. Fulle, H. F. Hamann and H. Hippler, *Phys. Chem. Chem. Phys.* **1999**, *1*, 2695-2702.

[98] K. Yamasaki, A. Watanabe, T. Kakuda and I. Tokue, *J. Phys. Chem. A* **2000**, *104*, 9081-9086.

[99] L. D'Ottone, D. Bauer, P. Campuzano-Jost, M. Fardy and A. J. Hynes, *Phys. Chem. Chem. Phys.* **2004**, *6*, 4276-4282.

[100] R. J. Shannon, S. Taylor, A. Goddard, M. A. Blitz and D. E. Heard, *Phys. Chem. Chem. Phys.* **2010**, *12*, 13511-13514.

[101] R. J. Shannon, R. L. Caravan, M. A. Blitz and D. E. Heard, *Phys. Chem. Chem. Phys.* **2014**, *16*, 3466-3478.

[102] I. J. Wysong, J. B. Jeffries and D. R. Crosley, *J. Chem. Phys.* **1990**, *92*, 5218-5222.

[103] J. B. Burkholder, A. Mellouki, R. Talukdar and A. R. Ravishankara, *Int. J. Chem. Kinet.* **1992**, *24*, 711-725.

[104] T. Yamada, P. H. Taylor, A. Goumri and P. Marshall, *J. Chem. Phys.* **2003**, *119*, 10600-10606.

[105] A. Febo, C. Perrino, M. Gherardi and R. Sparapani, *Envir. Sci. Technol.* **1995**, *29*, 2390-2395.

[106] O. Maass and P. G. Hiebert, *J. Am. Chem. Soc.* **1924**, *46*, 2693-2700.

[107] M. Baasandorj, D. K. Papanastasiou, R. K. Talukdar, A. S. Hasson and J. B. Burkholder, *Phys. Chem. Chem. Phys.* **2010**, *12*, 12101-12111.

[108] A. C. Egerton, W. Emte and G. J. Minkoff, *Discuss. Faraday Soc.* **1951**, *10*, 278-282.

[109] L. Messaadia, G. El Dib, A. Ferhati and A. Chakir, *Chem. Phys. Lett.* **2015**, *626*, 73-79.

[110] A. L. Holloway, J. Treacy, H. Sidebottom, A. Mellouki, V. Daële, G. Le Bras and I. Barnes, *J. Photochem. Photobiol. A* **2005**, *176*, 183-190.

[111] M. Inoue, Y. Arai, S. Saito and N. Suzuki, *J. Chem. Eng. Data* **1981**, *26*, 287-293.

[112] G. B. Taylor, *Ind. Eng. Chem.* **1925**, *17*, 633-635.

[113] R. Thévenet, A. Mellouki and G. Le Bras, *Int. J. Chem. Kinet.* **2000**, *32*, 676-685.

[114] C. Vovelle, A. Bonard, V. Daële and J. L. Delfau, *Phys. Chem. Chem. Phys.* **2001**, *3*, 4939-4945.

[115] T. J. Dillon, D. Holscher, V. Sivakumaran, A. Horowitz and J. N. Crowley, *Phys. Chem. Chem. Phys.* **2005**, *7*, 349-355.

[116] G. Goor, J. Glenneberg and S. Jacobi, in: B. Elvers (Ed.), *Ullmann's Encyclopedia of Industrial Chemistry*, **2012**, Vol. 18, Wiley-VCH Verlag GmbH & Co. KGaA: Weinheim, 393-427.

[117] A. Schiffman, D. D. Nelson and D. J. Nesbitt, *J. Chem. Phys.* **1993**, *98*, 6935-6946.

[118] A. Mellouki, S. Teton and G. Le Bras, *Int. J. Chem. Kinet.* **1995**, *27*, 791-805.

[119] H. Hippler, N. Krasteva, S. Nasterlack and F. Striebel, *J. Phys. Chem. A* **2006**, *110*, 6781-6788.

[120] J. Eble, *private communication*, **2017**.

[121] K. Glänzer and J. Troe, *Ber. Bunsenges. Phys. Chem.* **1974**, *78*, 71-76.

[122] J. N. Crowley, J. P. Burrows, G. K. Moortgat, G. Poulet and G. Le Bras, *Int. J. Chem. Kinet.* **1993**, *25*, 795-803.

[123] K. Bogumil, J. Orphal, T. Homann, S. Voigt, P. Spietz, O. C. Fleischmann, A. Vogel, M. Hartmann, H. Kromminga, H. Bovensmann, J. Frerick and J. P. Burrows, *J. Photochem. Photobiol. A* **2003**, *157*, 167-184.

[124] C. C. Addison, *Chem. Rev.* **1980**, *80*, 21-39.

[125] M. F. Merienne, A. Jenouvrier, B. Coquart and J. P. Lux, *J. Atmos. Chem.* **1997**, *27*, 219-232.

[126] M. Pfeifle, *private communication*, **2014**.

[127] U. R. Kunze and G. Schwedt, *Grundlagen der quantitativen Analyse*, **2009**, 6th Ed., Wiley-VCH: Weinheim.

[128] R. Atkinson, D. L. Baulch, R. A. Cox, J. N. Crowley, R. F. Hampson, R. G. Hynes, M. E. Jenkin, M. J. Rossi and J. Troe, *Atmos. Chem. Phys.* **2004**, *4*, 1461-1738.

[129] A. Saposchnikow, *Z. Phys. Chem.* **1904**, *49*, 697-708.

[130] A. Saposchnikow, *Z. Phys. Chem.* **1905**, *51*, 609-630.

[131] S. R. M. Ellis and J. M. Thwaites, *J. Appl. Chem.* **1957**, *7*, 152-160.

[132] S. A. Carl, T. Ingham, G. K. Moortgat and J. N. Crowley, *Chem. Phys. Lett.* **2001**, *341*, 93-98.

[133] M. Pfeifle, *Zur Rolle nichtthermischer Molekülpopulationen in der Kinetik atmosphärenchemischer Reaktionen*, Ph.D. thesis, Karlsruher Institut für Technologie, **2015**.

[134] J. Eble, *Kinetische Untersuchungen der Reaktionen von 2,5-Dimethylfuran und 2-Methylfuran mit Hydroxyladikalen*, Diploma thesis, Karlsruher Institut für Technologie, **2013**.

[135] A. Owen, *Kinetische Untersuchungen der Reaktion von 2-Methylfuran mit OH-Radikalen*, Bachelor thesis, Karlsruher Institut für Technologie, **2017**.

[136] M. Müller and U. Hübsch, in: B. Elvers (Ed.), *Ullmann's Encyclopedia of Industrial Chemistry*, **2012**, Vol. 11, Wiley-VCH Verlag GmbH & Co. KGaA: Weinheim, 305-308.

[137] T. A. Semelsberger, R. L. Borup and H. L. Greene, *J. Power Sources* **2006**, *156*, 497-511.

[138] C. Arcoumanis, C. Bae, R. Crookes and E. Kinoshita, *Fuel* **2008**, *87*, 1014-1030.

[139] M. Werner and G. Wachtmeister, *Motortech. Z.* **2010**, *71*, 540-542.

[140] S. H. Park and C. S. Lee, *Energy Convers. Manage.* **2014**, *86*, 848-863.

[141] R. D. Cook, D. F. Davidson and R. K. Hanson, *J. Phys. Chem. A* **2009**, *113*, 9974-9980.

[142] L. Nelson, O. Rattigan, R. Neavyn, H. Sidebottom, J. Treacy and O. J. Nielsen, *Int. J. Chem. Kinet.* **1990**, *22*, 1111-1126.

[143] P. A. Perry, R. Atkinson and J. N. Pitts, *J. Chem. Phys.* **1977**, *67*, 611-614.

[144] T. J. Wallington, R. Liu, P. Dagaut and M. J. Kurylo, *Int. J. Chem. Kinet.* **1988**, *20*, 41-49.

[145] A. Bonard, V. Daële, J. L. Delfau and C. Vovelle, *J. Phys. Chem. A* **2002**, *106*, 4384-4389.

[146] W. B. DeMore and K. D. Bayes, *J. Phys. Chem. A* **1999**, *103*, 2649-2654.

[147] F. P. Tully and A. T. Droege, *Int. J. Chem. Kinet.* **1987**, *19*, 251-259.

[148] R. S. Tranter and R. W. Walker, *Phys. Chem. Chem. Phys.* **2001**, *3*, 4722-4732.

[149] T. J. Wallington, J. M. Andino, L. M. Skewes, W. O. Siegl and S. M. Japar, *Int. J. Chem. Kinet.* **1989**, *21*, 993-1001.

[150] A. Bottoni, P. Della Casa and G. Poggi, *J. Mol. Struct.* **2001**, *542*, 123-137.

[151] F. Atadinç, C. Selçuki, L. Sari and V. Aviyente, *Phys. Chem. Chem. Phys.* **2002**, *4*, 1797-1806.

[152] J. Y. Wu, J. Y. Liu, Z. S. Li and C. C. Sun, *J. Chem. Phys.* **2003**, *118*, 10986-10995.

[153] A. M. El-Nahas, T. Uchimaru, M. Sugie, K. Tokuhashi and A. Sekiya, *J. Mol. Struct.* **2005**, *722*, 9-19.

[154] T. Ogura, A. Miyoshi and M. Koshi, *Phys. Chem. Chem. Phys.* **2007**, *9*, 5133-5142.

[155] C. Zavala-Oseguera, J. R. Alvaraz-Idaboy, G. Merino and A. Galano, *J. Phys. Chem. A* **2009**, *113*, 13913-13920.

[156] C. W. Zhou, J. M. Simmie and H. J. Curran, *Phys. Chem. Chem. Phys.* **2010**, *12*, 7221-7233.

[157] C. Bänsch, J. Kiecherer, M. Szöri and M. Olzmann, *J. Phys. Chem. A* **2013**, *117*, 8343-8351.

[158] M. Sakuth, T. Mensing, J. Schuler, W. Heitmann, G. Strehlke and D. Mayer, in: B. Elvers (Ed.), *Ullmann's Encyclopedia of Industrial Chemistry*, **2012**, Vol. 13, Wiley-VCH Verlag GmbH & Co. KGaA: Weinheim, 433-449.

[159] W. Chen, J. M. Lin, L. Reinhart and J. H. Weisburger, *Mutat. Res.-Fund. Mol. M.* **1993**, *287*, 227-233.

[160] M. Naito, C. Radcliffe, Y. Wada, T. Hoshino, X. Liu, M. Arai and M. Tamura, *J. Loss Prevent. Proc.* **2005**, *18*, 469-473.

[161] S. Di Tommaso, P. Rotureau, B. Sirjean, R. Fournet, W. Benaissa, P. Gruez and C. Adamo, *Process Saf. Prog.* **2013**, *33*, 64-69.

[162] M. M. Welzel, S. Schenk, M. Hau, H. K. Cammenga and H. Bothe, *J. Hazard. Mater.* **2000**, *72*, 1-9.

[163] B. Bailey, J. Eberhardt, S. Goguen and J. Erwin, *SAE Tech. Paper* **1997**, 972978.

[164] R. Anand and N. V. Mahalakshmi, *Proc. Inst. Mech. Eng. D J. Automob. Eng.* **2007**, *221*, 109-116.

[165] N. Kapilan, P. Mohanan and R. P. Reddy, *SAE Tech. Paper* **2008**, 2008-01-2466.

[166] C. Cinar, Ö. Can, F. Sahin and H. S. Yucesu, *Appl. Therm. Eng.* **2010**, *30*, 360-365.

[167] D. C. Rakopoulos, C. D. Rakopoulos, E. G. Giakoumis and A. M. Dimaratos, *Energy* **2012**, *43*, 214-224.

[168] A. Paul, P. K. Bose, R. Panua and D. Debroy, *J. Energy Inst.* **2015**, *88*, 1-10.

[169] M. Iranmanesh, J. P. Subrahmanyam and M. Babu, *SAE Tech. Paper* **2008**, 2008-01-1805.

[170] D. H. Qi, H. Chen, L. M. Geng and Y. Z. Bian, *Renew. Energ.* **2011**, *36*, 1252-1258.

[171] S. Sivalakshmi and T. Balusamy, *Fuel* **2013**, *106*, 106-110.

[172] S. Imtenan, H. H. Masjuki, M. Varman, I. M. R. Fattah, H. Sajjad and M. I. Arbab, *Energy Convers. Manage.* **2015**, *94*, 84-94.

[173] K. Yasunaga, F. Gillespie, J. M. Simmie, H. J. Curran, Y. Kuraguchi, H. Hoshikawa, M. Yamane and Y. Hidaka, *J. Phys. Chem. A* **2010**, *114*, 9098-9109.

[174] Y. Sakai, J. Herzler, M. Werler, C. Schulz and M. Fikri, *Proc. Comb. Inst.* **2017**, *36*, 195-202.

[175] L. S. Tran, J. Pieper, H. H. Carstensen, H. Zhao, I. Graf, Y. Ju, F. Qi and K. Kohse-Höinghaus, *Proc. Comb. Inst.* **2017**, *36*, 1165-1173.

[176] J. Eble, J. Kiecherer and M. Olzmann, *Z. Phys. Chem.* **2017**, *231*, 1603-1623.

[177] A. C. Lloyd, K. R. Darnall, A. M. Winer and J. N. Pitts, *Chem. Phys. Lett.* **1976**, *42*, 205-209.

[178] P. J. Bennett and J. A. Kerr, *J. Atmos. Chem.* **1989**, *8*, 87-94.

[179] P. J. Bennett and J. A. Kerr, *J. Atmos. Chem.* **1990**, *10*, 27-38.

[180] M. Semadeni, D. W. Stocker and J. A. Kerr, *J. Atmos. Chem.* **1992**, *16*, 79-93.

[181] L. Sandhiya, S. Ponnusamy and K. Senthilkumar, *RSC Adv.* **2016**, *6*, 81354-81363.

[182] M. Szöri, *private communication*, **2012**.

[183] J. Kiecherer, *private communication*, **2017**.

[184] J. Hetzler, *private communication*, **2014**.

[185] D. Stoye, in: B. Elvers (Ed.), *Ullmann's Encyclopedia of Industrial Chemistry*, **2012**, Vol. 33, Wiley-VCH Verlag GmbH & Co. KGaA: Weinheim, 620-688.

[186] R. Brück, P. Hirth, E. Jacob and W. Maus, in: R. Basshuysenvan and F. Schäfer (Eds.), *Handbuch Verbrennungsmotor*, **2015**, Springer Fachmedien: Wiesbaden, 1187-1194.

[187] J. Burger, M. Siegert, E. Ströfer and H. Hasse, *Fuel* **2010**, *89*, 3315-3319.

[188] M. Tuner, *SAE Tech. Paper* **2016**, 2016-01-0882.

[189] C. A. Daly, J. M. Simmie, P. Dagaut and M. Cathonnet, *Combust. Flame* **2001**, *125*, 1106-1117.

[190] H. J. Curran, E. M. Fisher, P. A. Glaude, N. M. Marinov, W. J. Pitz, C. K. Westbrook, D. W. Layton, P. F. Flynn, R. P. Durrett, A. O. zur Loye, O. C. Akinyemi and F. L. Dryer, *SAE Tech. Paper* **2001**, 2001-01-0653.

[191] V. Dias, X. Lories and J. Vandooren, *Combust. Sci. Technol.* **2010**, *182*, 350-364.

[192] L. Marrodán, E. Royo, A. Millera, R. Bilbao and M. U. Alzueta, *Energy Fuels* **2015**, *29*, 3507-3517.

[193] L. Marrodán, F. Monge, A. Millera, R. Bilbao and M. U. Alzueta, *Combust. Sci. Technol.* **2016**, *188*, 719-729.

[194] E. Hu, Z. Gao, Y. Liu, G. Yin and Z. Huang, *Fuel* **2017**, *189*, 350-357.

[195] W. Sun, G. Wang, S. Li, R. Zhang, B. Yang, J. Yang, Y. Li, C. K. Westbrook and C. K. Law, *Proc. Comb. Inst.* **2017**, *36*, 1269-1278.

[196] R. Atkinson and J. Arey, *Chem. Rev.* **2003**, *103*, 4605-4638.

[197] H. J. Curran, W. J. Pitz, C. K. Westbrook, P. Dagaut, J. C. Boettner and M. Cathonnet, *Int. J. Chem. Kinet.* **1998**, *30*, 229-241.

[198] A. T. Droege and F. P. Tully, *J. Phys. Chem.* **1986**, *90*, 1949-1954.

[199] N. Cohen, *Int. J. Chem. Kinet.* **1991**, *23*, 397-417.

[200] E. Porter, J. Wenger, J. Treacy, H. Sidebottom, A. Mellouki, S. Téton and G. Le Bras, *J. Phys. Chem. A* **1997**, *101*, 5770-5775.

[201] T. J. Wallington, M. D. Hurley, J. C. Ball, A. M. Straccia, J. Platz, L. K. Christensen, J. Sehested and O. J. Nielsen, *J. Phys. Chem. A* **1997**, *101*, 5302-5308.

[202] L. P. Thüner, I. Barnes, T. Maurer, C. G. Sauer and K. H. Becker, *Int. J. Chem. Kinet.* **1999**, *31*, 797-803.

[203] Y. R. Luo, *Handbook of Bond Dissociation Energies in Organic Compounds*, **2003**, CRC Press LLC: Boca Raton.

List of Publications

Articles in Reviewed Journals

- 'Reaction of Dimethyl Ether with Hydroxyl Radicals: Kinetic Isotope Effect and Prereactive Complex Formation',
  C. Bänsch, J. Kiecherer, M. Szöri and M. Olzmann, *J. Phys. Chem. A* **2013**, *117*, 8343-8351.
- 'Pyrolysis of ethanol: A shock-tube/TOF-MS and modeling study',
  J. Kiecherer, C. Bänsch, T. Bentz and M. Olzmann, *Proc. Comb. Inst.* **2015**, *35*, 465-472.
- 'Experimental study on the kinetics of the reaction of dimethoxymethane with hydroxyl radicals under autoignition-relevant conditions',
  C. Bänsch and M. Olzmann, *in preparation.*
- 'The kinetics of the reaction of diethyl ether with hydroxyl radicals and its isotope effect – a comprehensive experimental and theoretical study',
  C. Bänsch, J. Kiecherer, M. Szöri and M. Olzmann, *in preparation.*

Patents

- 'Hochdrucksättiger für korrosive und hochreine Flüssigkeiten',
  C. Bänsch, P. Hibomvschi and M. Olzmann, patent pending, appl. no. 102017123985.5, **2017**.

Conference Proceedings

- 'Thermal decomposition of ethanol – shock-tube study and kinetic modeling',
  J. Kiecherer, T. Bentz, C. Hüllemann, K. Blumenstock and M. Olzmann, *Proceedings of the European Combustion Meeting* **2011**, Cardiff (United Kingdom).
- 'Kinetic investigation of the reactions of 2,5-dimethylfuran and 2-methylfuran with hydroxyl radicals',
  J. Eble, C. Bänsch and M. Olzmann, *Proceedings of the European Combustion Meeting* **2015**, Budapest (Hungary).

Conference Distributions: Oral Presentations

- 'Zeitaufgelöste massenspektrometrische Untersuchung der Pyrolyse von Ethanol hinter reflektierten Stoßwellen',
  T. Bentz, J. Kiecherer, C. Hüllemann and M. Olzmann, *109. Hauptversammlung der Deutschen Bunsen-Gesellschaft für Physikalische Chemie e.V.* **2010**, Bielefeld (Germany).

- 'Isotope effect in the reactions of alkyl ethers with OH radicals', J. Kiecherer, C. Hüllemann, J. Hetzler and M. Olzmann, *111. Hauptversammlung der Deutschen Bunsen-Gesellschaft für Physikalische Chemie e.V.* **2012**, Leipzig (Germany).

- 'Initial steps in the pyrolysis and oxidation of dimethyl ether and diethyl ether', C. Hüllemann, J. Kiecherer, S. Faas, J. Hetzler, B. Derstroff and M. Olzmann, *3rd Annual Meeting COST CM0901 – Detailed Models for Cleaner Combustion* **2012**, Sofia (Bulgaria).

- 'Autoignition mechanisms and kinetics of diethyl ether-containing mixtures', C. Hüllemann, J. Kiecherer, B. Derstroff, S. Faas, J. Hetzler, T. Bentz and M. Olzmann, *International Seminar on Physicochemical-based Models for the Prediction of safety-relevant Ignition Processes, Research Group FOR 1447* **2013**, Tutzing (Germany).

- 'Pyrolysis of ethanol: A shock-tube/TOF-MS and modeling study', J. Kiecherer, C. Bänsch, T. Bentz and M. Olzmann, *35th International Symposium on Combustion* **2014**, San Francisco (United States of America).

- 'The reaction of diethyl ether with OH radicals: rate coefficient, kinetic isotope effect, and mechanism', C. Bänsch, J. Kiecherer, J. Hetzler, M. Szöri and M. Olzmann, *24th International Symposium on Gas Kinetics and Related Phenomena* **2016**, York (United Kingdom).

- 'Autoignition mechanisms and kinetics of diethyl ether-containing mixtures', J. Kiecherer, J. Eble, C. Bänsch, B. Derstroff, J. Hetzler and M. Olzmann, *International Workshop on Physicochemical-Based Models for the Prediction of Safety-Relevant Ignition Processes, Research Group FOR 1447* **2016**, Villa Vigoni, Loveno di Mennaggo (Italy).

- 'Sensitivität von Zündgrenzen gegenüber chemischer Kinetik', J. Kiecherer, J. Eble, C. Bänsch, B. Derstroff, J. Hetzler and M. Olzmann, *DECHEMA-Kolloquium "Kenngrößen für sicherheitstechnische Zündprozesse – Empirie oder vorausberechenbar?"* **2016**, Karlsruhe (Germany).

- 'Autoignition of diethyl ether: development of a mechanism and kinetic modeling', J. Eble, J. Kiecherer, C. Bänsch and M. Olzmann, *16th General Assembly of the German Bunsen Society for Physical Chemistry* **2017**, Kaiserslautern (Germany).

Conference Contributions: Poster Presentations

- 'Shock-tube study on the thermal decomposition of ethanol and its reaction with O_2', T. Bentz, J. Kiecherer, C. Hüllemann, K. Blumenstock and M. Olzmann, *33rd International Symposium on Combustion* **2010**, Beijing (China).

- 'Thermal decomposition of ethanol and its reaction with O_2 – Shock-tube study and kinetic modeling', J. Kiecherer, T. Bentz, C. Hüllemann, K. Blumenstock and M. Olzmann, *1st annual meeting, COST CM0901 – Detailed Models for Cleaner Combustion* **2010**, Nancy (France).

- 'The rate coefficient of the OH + HNO_3 reaction over a wide temperature and pressure range',
C. Bänsch, J. Crowley, T. J. Dillon, K. Dulitz, M. Olzmann, M. Pfeifle and M. Szöri, *23rd International Symposium on Gas Kinetics and Related Phenomena* **2014**, Szeged (Hungary).

- 'Kinetic investigation of the reactions of 2,5-dimethylfuran and 2-methylfuran with hydroxyl radicals',
J. Eble, C. Bänsch and M. Olzmann, *1st Annual Meeting COST Action SMARTCATs CM1404* **2015**, Thessaloniki (Greece).

- 'Kinetic Investigations of the Pyrolysis and Oxidation of Dimethoxymethane',
L. Golka, C. Bänsch, K. Wegner and M. Olzmann, *116th General Assembly of the German Bunsen Society for Physical Chemistry* **2017**, Kaiserslautern (Germany).